600种手绘多肉植物图谱
600 Colour Paintings of Succulent Plants

朱亮锋　编著

中国·武汉

图书在版编目（CIP）数据

600种手绘多肉植物图谱/朱亮锋编著. -- 武汉：华中科技大学出版社，2016.7
ISBN 978-7-5680-1768-8

Ⅰ.①600… Ⅱ.①朱… Ⅲ.①多浆植物—图谱 Ⅳ.①S682.33-64

中国版本图书馆CIP数据核字(2016)第091949号

600种手绘多肉植物图谱
600 Zhong Shouhui Duorou Zhiwu Tupu

朱亮锋 编著

出版发行：华中科技大学出版社（中国·武汉）	
地　　址：武汉市武昌珞喻路1037号（邮编：430074）	
出 版 人：阮海洪	
策划编辑：王　斌	责任监印：张贵君
责任编辑：王清瑢	装帧设计：百彤文化
校　　对：李用华	
印　　刷：雅昌文化（集团）有限公司	
开　　本：965mm×1270mm　1/16	
印　　张：19.5	
字　　数：300千字	
版　　次：2016年7月第1版　第1次印刷	
定　　价：318.00元（USD 63.99）	

投稿热线：(020)66636689　342855430@qq.com
本书若有印装质量问题，请向出版社营销中心调换
全国免费服务热线：400-6679-118 竭诚为您服务
版权所有　侵权必究

朱亮锋，教授，研究员，汉族，1937年1月出生于广州，1960年毕业于中山大学化学系，1964年在中国科学院华南植物研究所从事植物化学和植物资源学的研究，曾任华南植物研究所植物资源研究室副主任、主任，从1987年开始从事芦荟和中药材研究开发工作，曾任联合国教科文组织（UNESCO）亚洲及太平洋地区药用植物和芳香植物情报网络（APINMAP）联络员和中国国家中心负责人。1989年获国务院有突出贡献科学家津贴；1999年获国际精油和香料行业联盟（IFEAT）成就奖章。1999年退休。

Zhu Liangfeng, Professor, born on January 15, 1937 in Guangzhou, who graduated from the Department of Chemistry of Zhongshan University in 1960. He engaged in Phytochemistry and Plant Resource research at South China Institute of Botany, Academia Sinica from 1964 and started the research and development works of Aloe from 1987. He used to be the head of the department of plant resources, the liaison officer and the head of Chinese National Center of UNESCO Asian Pacific Information Network on Medicinal and Aromatic Plants (APINMAP). He also gained the Notable Contribution Scientist Allowance from State Council of the People's Republic of China in 1989 and the Achievements Medal from International Federation of Essential Oils and Aroma Trades (IFEAT) in 1999. Recently, he is the member of technical committee and the specially invited director of the Aloe Society of China.

吴萍，女，汉族，博士，1969年10月生于贵阳，1992年毕业于兰州大学生物系，现为中国科学院华南植物园副研究员。

Wu ping Ph.Dr. born in Guiyang City October 1969. Graduated from Department of Biology, Lanzhou University in 1992, Currently assistant Professor of South China Botanical Garden, Chinese Academy of Sciences.

王辰，男，汉族，1988年4月出生于兰州。2010年毕业于兰州大学化学化工学院，现于中国科学院华南植物园攻读博士学位。

Wang Chen, male, born in Lanzhou, April, 1988. Graduated from Department of Chemistry and Chemical Engineering, Lanzhou University in 2010. Currently as a Ph.D. candidate of South China Botanical Garden, Chinese Academy of Sciences.

前言

　　50多年的爱好和40多年的收集和积累，在各位同窗好友的帮助、鼓励和支持下，终于在退休后接近80的途中，如愿将600余种多肉植物彩色图谱呈现给大家。

　　在退休前借工作之便，偷闲到世界各地多肉植物的主要分布地区和各地区的植物园、植物公园拍照收集，有"仙人掌王国"之称的墨西哥、美国东南部多个州的国家公园和自然分布地区、加勒比海一些仙人掌分布岛屿等；百合科芦荟属、番杏科部分种属主要分布地区：南非、比勒陀尼亚、斯威士兰以及它们的植物园。另外还到过南半球的新西兰威灵顿植物园、澳大利亚墨尔本和悉尼植物园、东南亚印尼的茂密和巴厘岛植物园、新加坡植物园、法国巴黎植物园、摩纳哥植物园、加拿大温哥华室内植物园等。当年也走遍了我国各大植物园：北京香山植物园、南京中山植物园、昆明植物园、西双版纳植物园、厦门植物园、深圳仙湖植物园、武汉植物园、庐山植物园、西安植物园、贵阳植物园、上海龙华植物园和华南植物园等，前后共收集四千多张照片和有关资料，再经多年整理、筛选、汇总和查阅相关资料，最后临摹成画与大家见面。

　　既然是手绘的画，主要是将原始素材的每一种多肉植物通过照片、图片临摹加工成画，由于作者并非专业绘画之人，也并非植物分类学家、园艺栽培专家，对于确定每种多肉植物的命名和形态分布描述，在多名专家指点下还算见得人，但我估计错漏还是免不了的，敬请大家多多包涵。在本图谱中，由于时间和专业水平等问题不能把一些栽培种的形态和出处加以介绍。另外，由于参与本图谱（也可以说是画册）出版和制作的人员都不是专业从事植物分类学和园艺工作的，所以每种多肉植物画的说明只是针对原画本身的粗略介绍，缺乏专业水平，用词也不当，请多包涵，谢谢！

　　600余种手绘多肉植物图谱是通过多人参与，照片整理、资料收集、中英文简介、临摹过程的指点、每种图的扫描、出版等环节，少一个都不能完成，所以这本图谱不只是作者本人的，而是所有参与者的，在此对他们表示谢意。

PROLOGUE

Before my retirement, by taking the advantage of business trip, I had visited many botanical gardens, botanical parks, and natural distribution areas of succulent plant all around the world to collect photographs and pictures, including Mexico– 'the cactus kingdom'; natural distribution areas and national parks in southeast America; several islands of Caribbean Sea; South Africa and Swaziland, where are the major distribution areas of genus Aloe and several genera and species of family Aizoaceae; Wellington Botanical Garden in New Zealand; Melbourne and Sydney Botanical Gardens in Australia; Kebon Raya Bogor and Bali Botanical Garden; Singapore Botanical Garden; Paris Botanical Garden in France; Monaco Botanical Garden; Vancouver Botanical Garden in Canada; and most of the botanical gardens in China, such as Xiangshan Botanical Garden in Beijing, Zhongshan Botanical Garden in Nanjing, Kuming Botanical Garden, Xishuangbanna Botanical Garden, Xiamen Botanical Garden, Xianhu Botanical Garden in Shenzhen, Wuhan Botanical Garden, Lushan Botanical Garden, Xian Botanical Garden, Guiyang Botanical Garden, Longhua Botanical Garden in Shanghai and South China Botanical Garden in Guangzhou. Over 4000 photographs had been taken during these trips, and after years of selection, organization, summary and review, I have finally finished these drawings.

Since they are hand painted, all of the drawings are copies of the original photographs or pictures. People who have involved in the production and publication process of these illustrations, including myself, are neither professional painters, horticulture experts, nor phytotaxonomists. All the captions are simple, rough introductions of the specified illustrations, and you may find some inappropriate expressions between the lines. I sincerely wish that you readers could forgive my mistakes in the nomenclature, morphology, and distribution descriptions of these succulent plants. On account of time limit and short of professional skills, I am sorry that the morphology and distribution of some cultivars are not included in these illustrations.

Without the help from my dear friends and colleagues, these hand painted drawings of over 600 succulent plant species would not have been finished. Hereby I deeply thank the ones who have helped me with the photograph organization, data collection, Chinese and English introduction, painting direction, pictures scan, and publishing procedure.

目录 CONTENTS

仙人掌科 (Cactaceae)

白花有刺萼仙人掌 /3
花冠球 /3
连山一个变种 /4
糠枇岩牡丹 /4
黑岩牡丹 /5
三角岩牡丹 /5
岩牡丹 /6
勃氏岩牡丹 /7
由贵柱 /7
红花关节柱（由贵柱）花 /8
碧琉璃兜丸 /8
六棱兜 /9
兜丸 /9
瑞凤玉 /10
碧鸾凤玉 /10
鸾凤玉锦 /11
鸾凤玉（僧帽）/11
六棱鸾凤阁 /12
鸾凤阁 /12
三角鸾凤玉 /13
般若 /13
熊爪玉 /14
松露玉 /14
短柱布氏柱 /15
巨人仙人掌 /16
翁柱 /17
白闪 /17
吹雪 /18
鳞片恐龙角 /18
壮农 /19
红花象牙球 /19
金环触 /20
大祥冠 /20

小花粗刺顶花球 /21
粗刺顶花球 /21
长刺顶花球 /22
奇特球 /22
赫氏圆盘玉 /23
（艾氏）红花双重叶仙人掌 24
金鯱 /24
钩刺金鯱 /25
阔刺金鯱 /25
凌波金鯱 /26
武勇球（丛生鹿角柱）/26
训氏鹿角柱 /27
幻虾 /27
剑氏虾 /28
紫苑 /28
王将虾（长刺鹿角柱）/29
长刺虾 /29
春高楼 /30
三光球 /31
顶花虾 /32
硬刺鹿角柱 /32
太阳 /33
草木角 /33
大花虾 /34
稀刺鹿角柱 /34
微刺鹿角柱 /35
黑花青花虾 /35
美花鹿角柱 /36
篝火 /36
美花鹿角柱变种 /37
三刺虾 /37
青花虾 /38
威氏鹿角柱 /38
珠毛柱 /39
雪溪 /39

剑恋玉锦 /40
多棱玉 /40
千波万波锦 /41
五刺玉 /41
太刀岚 /42
龙舌玉（秋阵管）/42
缩玉锦 /43
缩玉 /43
仙人球 - 杂交栽培种 /44
金城 /44
沙漠柱 /44
年氏仙人球 /45
倒波形仙人球 /45
旺盛球 /46
红花毛花柱 /46
圆齿昙花 /47
多花昙花 /47
昙花花蕾 /48
昙花 /48
月世界 /49
小松球 /49
天龙 /50
秘鲁毛柱 /50
老乐柱 /51
小老乐 /51
短毛老乐 /52
老乐花 /52
短毛花壶柱 /53
白幻阁 /53
彩色强刺球 /54
琥头 /54
埃氏强刺球（江守）/55
刘穗玉 /55
红刺强刺球 /56
大虹 /56

恰氏强刺球 /57
箭状强刺球（变种）/57
箭状强刺球 /58
日之出（日出）/58
林氏强刺球 /59
赤风 /59
波氏强刺球 /60
长刺强刺球 /61
长刺荒鹫 /61
勇状球 /62
黄彩玉 /62
短刺强刺球 /63
钩刺强刺球（变种）/63
钩刺强刺球 /64
赤金龙（变种）/64
赤金龙（野外生长）/65
赤金龙（温室栽种）/65
钩刺球 /66
士童 /66
次大陆裸萼球 /67
翠晃冠（翠花冠）/67
绯花玉 /68
罗星球 /68
圣王丸 /69
卡迪那斯裸萼球 /69
天王球 /70
胭脂牡丹锦 /71
米氏裸萼球 /71
黑牡丹玉 /72
绯牡丹彩色裸萼球 /72
凤牡丹 /73
鸡冠绯牡丹 /73
珠红绯牡丹 /74
五彩绯牡丹 /74
祥云锦 /75

瑞云球（金色）/75
牡丹玉 /76
黑网孔裸萼球 /76
龙头 /77
新天地 /78
新天地 /79
沙氏裸萼球 /80
守殿玉 /80
分生裸萼球 /81
乌拉圭裸萼球 /81
威迪斯裸萼球 /82
金焰柱 /82
假昙花杂交种 /83
"巨犬"假昙花 /84
假昙花 /84
特美牡丹柱 /85
黄金纽属与仙人球属 - 杂交种 /86
量天尺 /86
光山 /87
黄裳 /87
黄裳（变种）/88
湘阳球 /88
凯南娜丽花球 /89
密刺丽花球（黄花）/89
华宝球 /90
阳盛球 /91
朱丽球 /91
红笠球 /92
红笠球（黄花）/92
布氏巨黄丸丛生变种 /93
青玉 /93
姬丽球（变种）/94
白檀 /94
白檀杂交种 /95
牡丹球 /95

桃轮球 /96
黑乳突球 /96
卡氏乳突球 /97
雪月花 /97
林氏乳突球 /98
金星 /98
柏氏乳突球 /99
月宫殿 /99
无刺乳突球 /100
黄仙玉 /100
帕拿利白玉仙 /101
白头花座球 /101
蓝云 /102
赫云 /102
层云 /103
爪刺花座球 /103
彩云 /104
姬云 /104
格氏花座球 /105
巴哈马花座球 /105
马坦萨斯花座球 /106
乱云 /106
华云 /107
碧云 /107
残雪 /108
红花大凤龙（勇凤）/108
智利多色球（豹头）/109
圆锥玉 /109
玉姬 /110
英冠玉（壮丽南国玉）/110
壮丽南国玉 /111
短尖刺南国玉 /111
拉氏南国玉 /112
眩美玉 /112
范氏南国玉 /113
大花南国玉（黄花）/113
南国玉-杂交栽培种 /114
未被鉴定南国玉 /114
南国玉 /115
毕氏团扇（红花团扇）/116
帝冠 /116
伯氏仙人掌 /117

松岚 /117
红点团扇 /118
长网孔仙人掌 /118
褐刺仙人掌 /119
仙人镜 /119
红团扇(淡褐刺仙人掌)/120
萨氏团扇 /121
鹤岑球 /122
圣云龙 /122
武烈丸 /123
红花天轮柱 /123
玻利维亚锦绣球 /124
拜氏南国玉 /124
海王球 /125
硬刺锦绣球 /125
赫氏南国玉 /126
黄翁 /126
红刺魔神 /127
黑云龙 /127
红色具刺南国玉 /128
金冠 /128
蛇状丝柱 /129
白眉塔 /129
木麒麟 /130
叶花仙人掌 /130
薇拉娜 /131
黄叶掌 /131
普氏叶掌杂交种 /132
乌夫人 /132
五月草 /133
红雀 /133
红叶掌杂交种 /134
洋娃娃马狄逊 /134
美国情人 /135
狮子 /135
伯氏毛柱 /136
直刺仙人掌 /136
网脉有翼柱 /137
黑龙 /137
真春黄菊子孙球 /138
珠节子孙球 /138
紫宝球 /139

优雅（仙人指杂交种）/140
黄金蟹（仙人指杂交种）/140
仙人指杂交种 /141
仙人指属杂交种 /141
园辨仙人指（锦上添花）/142
大花蛇鞭柱（大轮柱）/142
夜美人 /143
奇想球 /143
菊水 /144
豪猪刺新绿柱 /144
斯氏沟宝山 /145
近卫柱 /145
白林仙人掌 /146
大统领 /147
多色玉 /148
天昊 /148
图拉瘤玉球 /149
月章 /149
劳氏陀螺果 /150
精巧殿 /150
大花毛花柱 /151
施氏升龙 /152
鲛丽球 /153
瘤果鞭美玉恋 /154
卡氏花笠球 /154
软毛花笠球 /155
威斯汀花笠球 /155
红花威氏仙人掌 /156
蟹爪兰 /156
安第斯山的仙人掌 /157
安第斯山的凤梨科植物 /159

百合科 (Liliaceae)

具皮刺芦荟锦 /161
具皮刺芦荟 /161
疣突芦荟锦 /162
具皮刺芦荟锦 /162
亚非尼加芦荟（非洲芦荟）与好望角芦荟杂交种 /163
阿非利加芦荟与马洛夫芦荟杂交种 163
阿非利加芦荟 /164
相似芦荟 /165

巴伯芦荟 /165
巴特尔芦荟 /166
巴哈那芦荟 /166
博伊尔芦荟 /167
短叶芦荟 /167
红刺芦荟 /168
烛台芦荟 /168
栗褐芦荟 /169
睫毛芦荟 /169
康普顿芦荟 /170
簇叶芦荟 /170
圆锥芦荟 /171
大肚芦荟 /171
隐柄芦荟 /172
达本洛里斯芦荟 /172
威氏芦荟 /173
远距芦荟 /173
多花序芦荟 /174
戴尔芦荟 /174
短叶芦荟锦 /175
高芦荟 /175
好望角芦荟与马洛夫芦荟杂交种 /176
好望角芦荟 /177
粉绿芦荟 /178
球芽芦荟 /178
哈迪芦荟 /179
艳丽芦荟 /179
伊比戴芦荟 /180
喀来斯芦荟（老树）/180
喀来斯芦荟 /181
红线芦荟 /181
海滨芦荟 /182
长苞芦荟 /182
变黄色芦荟 /183
水晶芦荟 /183
斑痕芦荟 /184
皂素芦荟 /184
马洛夫芦荟与好望角芦荟杂交种 /185
马洛夫芦荟 /186
马洛夫芦荟 /187
黑刺芦荟 /187
易变芦荟 /188

佩格勒芦荟 /189
希帕尔芦荟 /190
巨箭筒芦荟 /191
巨箭筒芦荟（比兰斯芦荟）/191
多齿芦荟与红线荟杂交种 /192
扇状芦荟 /192
多齿芦荟 /193
多叶芦荟 /193
比勒陀尼亚芦荟 /194
赖茨芦荟 /195
喜石芦荟 /196
苏拿利芦荟 /197
索潘斯伯格芦荟 /197
奇丽芦荟与开普敦芦荟杂交种 /198
穗花芦荟 /198
珊瑚芦荟 /199
多浆芦荟 /199
开卷芦荟 /200
拉思卡芦荟 /201
沙丘芦荟 /201
钦科罗斯芦荟 /202
斑叶芦荟 /202
瓦奥姆比芦荟 /203
福尔肯芦荟与好望角芦荟杂交种 /204
福尔肯芦荟 /205
费雷赫德芦荟 /206
威肯斯芦荟 /207
具皮刺芦荟杂交种 /207
薄叶芦荟 /208
未鉴定芦荟一种 /209
重叶芦荟 /209
卧牛 /210
纯叶牛胭锦 /211
青龙刀 /211
阔叶青龙刀 /212
阔叶青龙刀锦 /212
短叶牛胭 /213
斑叶牛利 /213
大叶青龙刀 /214
金城 /214
水晶掌 /215
雅致十二卷 /215

雅致十二卷 /215
霜鹤 /216
琉璃殿锦 /216
瑞鹤 /217
霜百合 /217
翡翠十二卷 /218
五十之塔锦 /219

景天科 (Crassulaceae)

天章 /221
碎叶天章 /221
筒形百合莲花掌 /222
包叶莲花掌 /222
卷叶莲花掌 /223
长叶莲花掌 /223
扭叶莲花掌 /224
黑法师 /224
黑法师 /225
黑法师 /226
垂叶黑法师 /226
黑法师 /227
红叶黑法师 /228
尖叶黑法师 /228
黑法师 /229
山地玫瑰 /230
山地玫瑰 /231
红边莲花掌 /232
明镜 /232
莲花掌 /233
皱叶红覆轮 /234
红覆轮锦 /234
波浪叶银波锦 /235
红边银波锦 /235
银波锦 /236
紫叶银波锦 /236
火焰银波锦 /237
筒叶青锁龙 /237
星都 /238
红花月 /238
红花月 /239
玉春 /239

火祭 /240
红背青锁龙 /240
头状青锁龙 /241
塔形青锁龙 /241
彩巴 /242
神刀 /242
神刀 /243
乔木状青锁龙 /243
圆叶玉春 /244
圆叶状青锁龙 /244
竹节青锁龙 /245
白粉草 /245
雪莲 /246
紫红阔叶石莲 /246
黄玉石莲花 /247
艳红石莲（东云红叶）/247
吉娃莲 /248
优雅石莲花 /248
阔叶石莲花 /249
小囊状石莲花 /249
吉毕紫石莲 /250
铁石莲花 /250
淡云 /251
短柄石莲 /251
尖红毛石莲 /252
红石莲 /252
绒毛石莲 /253
扇贝石莲花 /253
特玉莲 /254
劳氏石莲花 /254
尖红石莲花 /255
长序石莲花 /255
匙叶石莲 /256
红花石莲 /256
灰白石莲花 /257
红边匙叶石莲 /257
红边翡翠石莲花 /258
尖红绒毛石莲 /258
紫红大石莲 /259
石莲花和青锁龙杂交种 /259
大叶紫石莲 /260
醉美人 /260

红边棒槌玉莲 /261
黄玉莲 /261
艳美人 /262
风车草和石莲花杂交种 /262
金叶仙女之舞 /263
大叶落地生根 /263
矾莲 /264
彩唐印 /264
长叶彩唐印 /265
铲叶唐印 /265
家种彩唐印 /266
扭叶彩唐印 /266
红边耳叶唐印 /267
洋吊钟 /267
桃美人 /268
多叶厚叶草 /268
红艳星美人 /269
塔形厚叶莲 /269
景天美人 /270
黄丽 /270
莲座光亮景天 /271
高加索景天 /271
光亮景天 /272
小妞妞 /272
垂叶观音莲 /273
长叶观音莲 /274
观音莲 /275
球状观音莲（菠萝球观音莲）275

番杏科 (Aizoaceae)

粒状菱鲛 /277
马哈比菱鲛 - 长生草 /277
莲座菱鲛 /278
威勒菱鲛 /278
马哈比菱鲛 /279
具毛菱鲛 /279
红菱鲛 /279
烛台肉堆花 /280
少将 /280
天使肉堆花 /280
碧玉肉堆花 /281

翠光玉 /281
小球肉堆花 /281
紫花肉堆花 /281
红花肉堆花 /281
橙黄花棒叶花 /282
白花棒叶花 /282
白花棒叶花 /283
棒叶花（橙红花）/283
百花光亮棒叶花 /283
紫花花棒叶花 /283
浅黄花光亮棒叶花 /283
红花日中花 /284
弯刀日中花 /284
日中花 /285
日轮玉 /286
橙黄花石生花 /286
黄花荒头 /287
红橄榄变种 /287
斑绿石生花 /287
红橄榄 /287
青玉 /287
红花石头 /287
红橄榄变种 /288
紫勋（红大内玉）/288
莫高妮石生花 /288
假截形石生花 /288
红大内玉 /288
富贵玉 /289
白花石生花 /289
黄花石生花 /289
福来玉 /289
留碟玉 /289
黄花石头 /289
蜡质石生花 /290
微纹玉 /290
黄花微纹玉 /290
紫翠玉 /290
巧克力石头 /290
红花石头变种 /290
绿玉 /291
金王花石生花 /291
疣突石生花变种 /291

波路氏对叶花 /292
亲鸾 /292
绿帝玉 /293
白帝玉 /293
紫帝玉（变种）/293
绿帝玉变种 /293
灰帝玉 /293
奈尔对叶花 /294
对叶肉堆花 /294
青鸾 /294
红花对叶花 /294
快刀乱麻 /295
尼尔快刀乱麻 /295
尼尔快刀乱麻 /296
斧状快刀乱麻 /296
斧状叶快刀乱麻 /297
快刀乱麻 /297
长距天女变种 /298
长距天女（红色）/298
长距天女变种 /299
白点天女 /299
长距天女 /299
红花天女 /299

仙人掌科 (Cactaceae)

仙人掌科植物是一个非常庞大的家族，绝大多数种属为肉质植物，按前期的植物分类研究被归类为233个属、670种，但尚未能把所有杂交、变种归纳进去。仙人掌科植物由于种与种之间、属与属之间不易划分，故还在不断归并和分类、变化不断，至1993年仙人掌科又被合并为98属，现在看来这个大家族还在不断演变中。

大多数仙人掌植物随着环境变化，其形态不断适应，最明显就是其叶在干旱恶劣环境中已经变成针刺状和长短不一绒毛状，而他的茎则变成肥大肉质、环状、筒状、柱状等形态，使之形成与其他植物不同的外貌。仙人掌科千姿百态、变幻神奇的植株为世人所喜爱，故仙人掌已成为各类温室必备的盆栽观赏花卉。

Cactaceae is a huge family which could be subdivided into 233 genera and 670 species according to previous taxonomy studies. However, all of the hybrids and variants are not included. Due to the difficulty in dividing the plants of family Cactaceae between species and species, genera and genera, the merging and classification of them have not been finished yet. The Cactaceae has been merged into 98 genera in 1993, it seems that this huge family is still evolving.

Most of the Cactaceae plants could constantly adapt to the changes of environment. As a result of growing in drought conditions, their leaves are usually needle-like, stem fleshy, succulent, circular, cannular or columnar. Their diverse and particular morphologies are deeply favored by people, and make them become the indispensable potted ornamental flowers in greenhouse.

白花有刺萼仙人掌
Acanthocalycium spiniflorum (K. Schum.) Backeb.

植株球形，绿色，具 16~18 条较阔直棱，周刺 6~10 条，较细小，中刺无或 1 根，黑色。花顶生，白色，花心黄色。原产阿根廷。

Whole plants spherical, green; 16 to 18 ribs; 6 to 10 small outer spines, 0 to 1 black central spines; flowers white, and acrogenous; native to Argentina.

花冠球
Acanthocalycium spiniflorum (K. Schum.) Backeb.

株高约 20cm，茎粗约 13cm，生有 15~20 条棱，周刺 6~10 根，中刺 1~4 根。花喇叭型，白色或粉红色至淡紫色。分布于阿根廷北部一个不大区域，为较稀少种类。

A rare species; 20 cm tall, stems diameter 13 cm; 15 to 20 ribs; 6 to 10 outer spines, 1 to 4 central spines; flowers white or pink to purple, trumpet shaped ; distribute in limited areas of northern Agentina.

连山一个变种
Ariocarpus fissuratus (Engelm.) K.Schum.

植株疣突上部扁平，中间有条纵向白色绒毛，植株浅灰色褐色，花红色。原产美国德克萨斯州。

Whole plants flat on top, pale greyish brown; flowers red; native to Texas, America.

糠秕岩牡丹
Ariocarpus furfuraceus Frič

植株灰绿色，疣突平滑肥厚，植株中间具白色绒毛，花淡红色。原产墨西哥。

Whole plants greyish green with smooth, fleshy warts; flowers pale red; native to Mexico.

黑岩牡丹
Ariocarpus kotschoubeyanus (Lem.) K. Schum.

与一般仙人掌科植物形态有些不同，看起来更像芦荟或龙舌兰科植物。植株扁平，阔 7 cm，疣突呈三角形，深绿色，生有带绒毛的褶皱，花淡紫色或粉红色，花宽 1.5~2.5 cm，相当好看。原产墨西哥。

Quite different in mophology when comparing to other plants in family Cactaceae, it looks more like Aloe or the plants in Agavaceae; whole plant flat, 7 cm wide with dark green, triangular warts; flowers pale purple or pink, 1.5 to 2.5 cm wide, which is quite lovely; native to Mexico.

三角岩牡丹
Ariocarpus retusus subsp. trigonus (F.A.C.Weber) E.F. Anderson & W.A.Fitz Maur.

植株扁，三角形，疣突灰绿色，末端钝，花呈紫粉红色，原产墨西哥。
Whole plant flat and triangle shaped; flowers purple pink; native to Mexico.

岩牡丹
Ariocarpus retusus Scheidw. (red)

岩牡丹
Ariocarpus retusus Scheidw. (white)

花为白绿色。广布于墨西哥的圣路易斯波托西、科阿韦拉和新莱昂州地区。
Flowers white-green; widely distribute in San Luis Potosi, Coahuila and Noevo Leon, Mexico.

勃氏岩牡丹
Ariocarpus scaphirostris Boed.

亦称龙角岩牡丹，株宽 3~7 cm，疣突呈灰绿色，顶端扁平无褶皱，末端平钝，花粉紫色。原产墨西哥。

Whole plants 3 to 7 cm wide; with greyish green warts; top flat and smooth; flowers pink-purple; native to Mexico.

由贵柱
Arrojadoa rhodantha (Gürke) Britton & Rose

植株直立，有时也呈蔓生和平卧，分枝呈深绿色，长 0.4~2 m，茎粗 2~4 cm，具 9~14 条较低的棱，生有 20 根周刺，中刺 5~6 根，每年可从长达 3 cm 的红色花座里生出紫红色花。花长 3~3.5 cm。原产南美巴西。

Whole plants erect, sometimes sprawling; branched, 0.4 to 2 m long, stem diameter 2 to 4 cm; 9 to 14 ribs; 20 outer spines, 5 to 6 central spines; purple-red flowers grow from a red, 3 cm long cephalium every year; native to Brazil.

红花关节柱（由贵柱）花
Arrojadoa rhodantha (Gürke) Britton & Rose

植株柱形，生有 9~14 条棱，花粉红色或鲜红，筒形。原产巴西。

Whole plants terete; 9 to 14 ribs; tubular flowers pink or bright red; native to Brazil.

碧琉璃兜丸
Astrophytum asterias var. nudum 'Hybrid'

5 条非常扁平的棱，整个植株没有本属的特点。白色星点。全株碧绿色十分为栽培者喜爱。

Whole plants aquamarine, with 5 flat ribs; deeply favored by plants enthusiasts.

兜丸
Astrophytum asterias (Zucc.) Lem.

植株扁球形，单生，球径8~10 cm，青绿色，具8条宽阔低平的棱，白色绒毛小星点布满整个球体，刺座具白色至淡褐色绒毛，但无刺，花顶生，明亮黄色，花心红棕色，漏斗状。原产美国德克萨斯州和墨西哥北部。

Whole plants flat spherical, solitary and blue-green; diameter 8 to 10 cm; 8 broad ribs; white to pale brown tomentum grow on spineless areole; flowers bight yellow and acrogenous. native to Texas, America and north Mexico.

六棱兜
Astrophytum asterias (Zucc.) Lem.

花大，为黄色。为兜丸的一个变种。分布于美国新墨西哥州和得克萨斯南部。
Flowers large, yellow. A cultivar of Astrophytum asterias. Distributed in New Mexico and southern Texas, the United States.

瑞凤玉
Astrophytum capricorne (A. Dietr.) Britton & Rose

植株筒形，花为深黄色，花期在夏季。分布于墨西哥北部。
Whole plants terete, flowers deep yellow, florescence is in summer. Distributed in northern Mexico.

碧鸾凤玉
Astrophytum myriostigma 'Variegata'

植株5条粗棱，无白色斑点，碧绿色，为鸾凤玉变种。
A variant of Astrophytum myriostigma; 5 thick ribs; whole plants aquamarine.

鸾凤玉锦
Astrophytum myriostigma 'Variegata'

花为黄色，花期为 5~8 月，分布于墨西哥中部到南部的高地。

Flowers yellow, florescence is from May to August. Widespread but scattered in the northern and central highlands of Mexico.

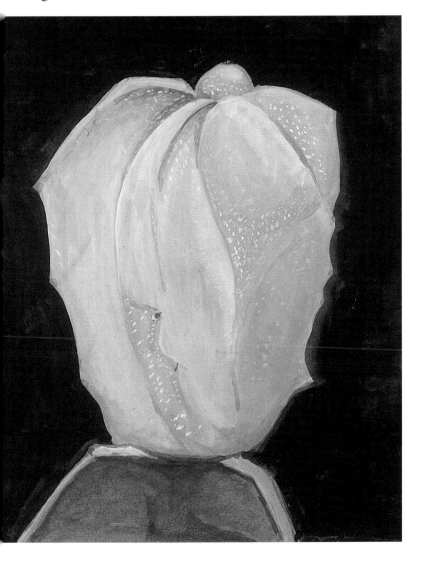

鸾凤玉（僧帽）
Astrophytum myriostigma Lem.

植株球形或略长，5 条肥厚低棱，棱缘上长有褐色无刺刺座，绒毛球十分密集，花黄色或红色，顶生，原产墨西哥，有很多变种，主要是棱数目变化。

Whole plant spherical or spheroidal with 5 ribs, brown spineless areole grow on the rib margin; acrogenous flowers yellow or red; native to Mexico.

六棱鸾凤阁
Astrophytum myriostigma var. columnarie 'Hexagemus'

植株筒形，具 6 条较粗的棱，棱缘无刺，为鸾凤玉的变种。

A variant of Astrophytum myriostigma; whole plant terete with 6 thick spineless ribs.

鸾凤阁
Astrophytum myriostigma var. columnarie Frič

柱状，花为黄色，花期为 5~8 月。分布于墨西哥中部到南部的高地。

whole plant terete; Flowers yellow, florescence is from May to August. Widespread but scattered in the northern and central highlands of Mexico.

般若
Astrophytum ornatum Lem (DC.) Britton & Rose

花为淡黄色,花期在春夏季,刺为金黄色。分布于墨西哥的伊达尔戈和克雷塔罗州。

Flowers pale yellow, florescence is from spring to summer; spines golden. Distributed in Hidalgo to Queretaro, Mexico.

三角鸾凤玉
Astrophytum myriostigma var. nudum 'Tricornis'

植株有三角肥大棱,长满白星点,无刺,花淡黄色,顶生,为鸾凤玉变种。

A variant of Astrophytum myriostigma; triangular fleshy ribs, covered with white terete; spineless; flowers pale yellow and acrogenous.

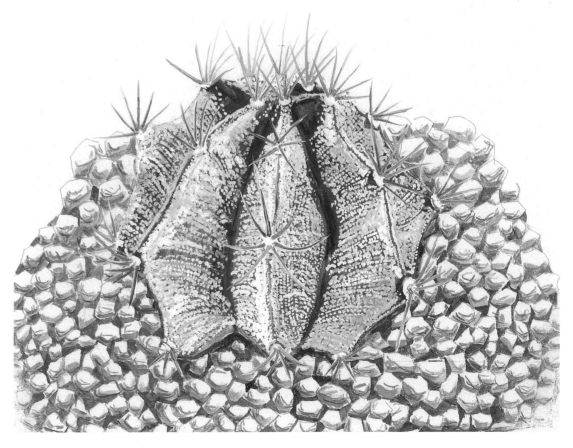

熊爪玉
Austrocactus patagonicus (Weber Backeb ex Speg.) Hosseus

植株高达50 cm，茎粗8 cm，生有9~12条带有疣突的棱，周刺6~16根，白色，刺尖呈褐色，1~2 cm长，中刺1~4根，长达4 cm，呈钩状。花白色或粉白色。原产阿根廷。

Whole plant 50 cm tall, stems diameter 8 cm; 9 to 12 ribs with warts; 6 to 16 white, 1 to 2 cm long outer spines with brown spine apex, 1 to 4 hooked, 4 cm long central spines;; flowers white or pink-white; native to Argentina.

松露玉
Blossfeldia liliputana Werderm.

株幅0.4~1 cm，是小型种类，花白色，密生。原产地为阿根廷北部，玻利维亚，是著名的小型种类仙人掌。

A small species; 0.4 to 1 cm wide; flowers white and dense; native to north Argentina, Bolivia.

短柱布氏柱
Buiningia brevicylindrica Buining

初生植株为球形，成长后渐变为短柱形，从基部分枝，高 30 cm，茎粗 17 cm，刺座上 18 条弓形棱，周刺 7 根约 2 cm 长，中刺 4 根，花座为布满白色棉毛及金黄色刚毛状刺，花奶油色，筒状，长 3.2 cm。原产巴西，不耐寒，种植控制在 10 ℃ 以上。

Newborn plants spherical, gradually turn into short terete; branched from stems base; 30 cm tall, stems diameter 17 cm; 18 arched ribs grow on areole; 7 outer spines up to 2 cm long, 4 central spines; white tomentum and golden bristle-like spines cover the cephalium; flowers cream white and tubular, up to 3.2 cm long; native to Brazil; intolerance to cold.

巨人仙人掌
Carnegiea gigantea (Engelm.) Britton & Rose

高大柱状，高达 52 英尺，花为白色，花期在春末。分布于美国亚利桑那州南部、加利福尼亚州南部和墨西哥的索诺拉州。

Treelike, growing to a height of fifty-two feet, flowers white, florescence is in the late spring. Distributed in Sonora, Mexico and southern Arizona, southern California, the United States.

翁柱
Cephalocereus senilis (Haw.) Pfeiff.

一种高大柱状仙人掌，高达 15 米，多不分枝，成年植株披满白色细长的绒毛，花黄白色，夜间开放。原产墨西哥。

Whole plants tall and big, up to 15 m high, usually not branched; flowers yellowish white, nocturnal; native to Mexico.

白闪
Cleistocactus jujuyensis (Backeb.) Backeb.

植株直立单生，成长后会不断从基部分枝成群，鲜绿色；全株被银白色细短刚毛状周刺所包裹。茎高约 1 m，具 22~26 条低棱，花管状，红色，管长约 7~8 cm，侧开。原产玻利维亚。

Whole plants bight green, erect and solitary, covered with short, white bristle like spines; branch from the base after mature; stems up to 1 m tall, 22 to 26 ribs; flowers red and tubular; native to Bolivia.

吹雪
Cleistocactus straussii Backeb.

植株直立，多分枝成群，高约 1 m，绿色，密布白色毛状细刚刺。花管状，侧生，红色，夏季开花，花管长 7~8 cm，花管为细长绒毛所包裹。原产玻利维亚和阿根廷。

Whole plants erect, usually branched, up to 1 m tall, green, covered with white, bristle like spines; flowers red and tubular, florescence is in summer; native to Bolivia and Argentina.

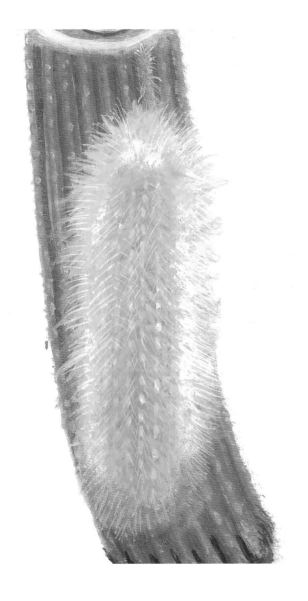

鳞片恐龙角
Corryocactus squarrosus (Vaupel) Hutchison

植株柱形，黄绿色或绿色，具 8~10 条直棱。周刺 8~12 根白色，中刺 1~3 根，褐色和黑色。花顶生黄色。原产玻利维亚、秘鲁等。

Whole plants terete, yellowish green or green; 8 to10 straight ribs; 8 to 12 white outer spines. 1 to 3 brown and black central spines; flowers yellow and acrogenous; native to Bolivia and Peru.

壮农
Coryphantha calipensis Bravo ex S. Arias Montes, U. Guzman Cruz & S. Gama Lopez

植株高 9 cm，株幅 5~8 cm，分枝，茎灰绿色或橄榄绿，疣突 3 cm，周刺 10~16 根，长 1~5 cm，中刺只有一根，长 1.5 cm。花黄色。原产墨西哥。

Whole plants 9 cm tall, 5 to 8 cm wide; branched; stems greyish green or olive; 10 to 16 outer spines, about 1 to 5 cm long, 1 central spine up to 1.5 cm long; flowers yellow; native to Mexico.

红花象牙球
Coryphantha elephantidens (Lem.) Lem.

植株深绿色有光泽球形，株高 14 cm，株幅 20 cm，基部长有明显球状疣突，疣突间长有白色棉状绒毛，周刺 5~8 根，青绿色，长 2 cm，无中刺，花顶开，粉红或朱红，原产墨西哥。

Whole plant dark green, glossy, and spherical; 14 cm high, 20 cm wide; spherical warts grow at base, white tomentum grow between the warts; 5 to 8 pale green, 2 cm long outer spines, no central spines; flowers pink or vermilion, acrogenous; native to Mexico.

金环触
Coryphantha pallida Britton & Rose

花为淡黄色,花期在 7~8 月。分布于墨西哥的普埃布拉。

Flowers light yellow, florescence is from July to August. Distributed in Puebla, Mexico.

大祥冠
Coryphantha poselgeriana (A. Dietr.) Britton & Rose

花丛生于植物顶部,紫红色。分布于墨西哥科阿韦拉州邻近萨尔蒂约地区。

Flowers clumping at the top of the plant, purple red. Distributed in Coahuila near Saltillo, Mexico.

小花粗刺顶花球
Coryphantha robustispina (Schott ex Engelm.) Britton & Rose

植株球形，黄绿色或绿色，刺座长于疣突上，周刺 7~9 根，褐黄色，中刺 1~3 根，红褐色，花顶生，黄色。原产美国南部和墨西哥北部。

Whole plants spherical, yellowish green or green; areole grow on warts; 7 to 9 brownish yellow outer spines, 1 to 3 brownish red central spines; flowers yellow and arogenous; native to southern U.S. and northern Mexico.

粗刺顶花球
Coryphantha robustispina (Schott ex Engelm.) Britton & Rose

花为金黄色到浅黄色。分布于美国亚利桑那州南部、新墨西哥州、得克萨斯州和墨西哥索诺拉州、奇瓦瓦州。

Flowers golden yellow to pale yellow. Distributed in southern Arizona, New Mexico, and Texas, the United States; Sonora and Chihuahua, Mexico.

长刺顶花球
Coryphantha werdermannii Boed.

花开于植株顶部，花大为橙色，顶部有细密长刺。主要分布于墨西哥。

Flowers large, orange, growing from the top of the plant with dense spines. Distributed in Mexico.

奇特球
Discocactus horstii Buining & Brederoo

植株单生，扁球形至圆球形，株高 2~3 cm，绿色至褐绿色。具 15~22 条较深纵向直棱；刺座长在棱缘上并有白色的绒毛团，具 8~10 枚灰白色且向内弯曲周刺。顶部具白色垫毛花座，高可达 1~2 cm，老植株会更高些；花较大，白色，漏斗状，夜间开花，有香味。原产巴西东部。

Whole plants solitary, flat spherical or spherical, 2 to 3 cm tall, green or brownish green; 15 to 22 straight ribs; areole covered with white tomentum and grow on rib margin; 8 to 10 greyish white outer spines curve inward; white cephalium grow on top, 1 to 2 cm high; funnel shaped, aromatic flowers are big and white, usually bloom at night; natvie to eastern Brazil.

赫氏圆盘玉
Discocactus horstii Buining et Brederoo

植株扁平球形，具最多达22条棱，长有8~10根短刺，呈褐色或灰白色，花顶生于一团白色绒毛。原产巴西。

Whole plants spherical, 22 ribs; 8 to 10 short, brown or greyish white spines; acrogenous flowers grow in white tomentum; native to Brazil.

（艾氏）红花双重叶仙人掌
Disocactus eichlamii (Weing.) Britton & Rose

植株长有细齿状的凹陷的棱，株幅宽 3~5 cm。花朱红色故亦称红花双重叶仙人掌。原产南美洲危地马拉和洪都拉斯。

Whole plants 3 to 5 cm wide; flowers vermilion; native to Guatemala and Honduras.

金琥
Echinocactus grusonii Hildm.

花小，为褐色，生长于植株顶部呈环状。分布于墨西哥。观赏价值颇高。

Flowers small and brown, appeared like a crown around the top of the plant. Distributed in Mexico.

阔刺金鯱
Echinocactus platyacanthus Link & Otto

花为黄色。分布于墨西哥中部和中北部的地区。

Flowers yellow. Distributed in a wide area of central and central northern Mexico.

钩刺金鯱
Echinocactus johnsonii Parry ex Engelm.

植株倒卵形，具 12~14 条浅棱，周刺 8~12 根，白色，中刺 2~3 根，灰白色。刺尖带钩，顶部根特长 5~6 cm。花顶生，红色，管状。原产墨西哥。

Whole plant obovate, 12 to 14 shallow arrises; flowers red, tube shaped, and apical; native to Mexico.

凌波金鯱
Echinocactus texensis Hopffer

花为肉色到紫罗兰色，丛生于植株顶部。分布于美国新墨西哥州东南部和得克萨斯州。

Acrogenous flowers cluster on top, colors ranging from salmon to violet. Distributed in Texas and southeastern New Mexico, the United States.

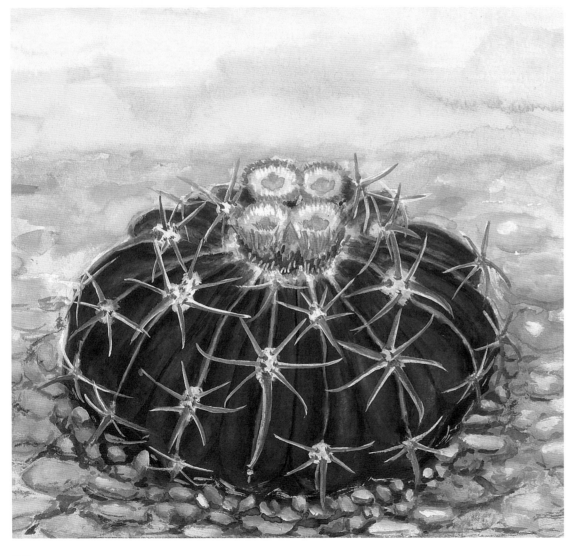

武勇球（丛生鹿角柱）
Echinocereus engelmannii (Parry ex Engelm.) Lem.

植株分枝，枝杈 4~6 cm，具 10~14 条略微隆起的棱，边缘刺约 10~12 根，中刺 2~6 根，长约 7 cm，刺的颜色多样，从淡黄至褐色；花淡紫色。原产美国西南各州和墨西哥北部。

Whole plants branched; 10 to 14 ribs; 10 to 12 outer spines, 2 to 6 central spines up to 7 cm long, the color of spines range from pale yellow to brown; flowers pale purple; native to southwest America and north Meixco.

训氏鹿角柱
Echinocereus fendleri (Engelm.) Sencke ex J.N. Haage

植株粗 4~8 cm，具 9~12 条棱，周刺 2~7 根，长 2.9 cm。花紫丁香色，花心颜色较深。分布于美国亚利桑那州中部和东部、科罗拉多州的南部、新墨西哥州、得克萨斯州的西部和墨西哥的奇瓦瓦州。

Whole plant diameter 4 to 8 cm, 9 to 12 ribs; flowers lilac; distributed in central and eastern Arizona, southern Colorado, New Mexico, western Texas, the United States; and Mexico in adjacent Chihuahua State.

幻虾
Echinocereus ferreianus H.E. Gates

植株球形，灰绿色，具 11~14 条较阔的直棱，周刺 8~14 根，中刺 4~7 根，长可达 10 cm，黑色或深褐色，花紫丁香色。原产墨西哥。幻虾的刺坚硬十分美丽，为栽培者喜爱。

Whole plants spherical, greyish green; 11 to 14 straight ribs; 8 to 14 outer spines, 4 to 7 black or dark brown, 10 cm long central spines; flowers lilac; native to Mexico; spines stiff and beautiful, favored by plant enthusiasts.

剑氏虾
Echinocereus ferreirianus var. lindsayi (J. Meyrán) N.P. Taylor

植株球形，绿色或灰绿色，具 11~14 条棱，周刺 8~14 根，中刺 4~7 根，长约 10 cm，棕黑色或褐色，花呈紫丁香色，颈喉为红色。原产墨西哥。

Whole plants spherical, green or greyish green; 11 to 14 ribs; 8 to 14 outer spines, 4 to 7 black or brown, 10 cm long central spines; flowers lilac; native to Mexico.

紫苑
Echinocereus ledingii Peebles

植株易分枝，花为深紫罗兰色。分布于美国亚利桑那州东南部。

Whole plants branched, flowers deep violet. Distributed in southeastern Arizona, the United States.

王将虾（长刺鹿角柱）
Echinocereus longisetus (Engelm.) Lem.

花大，紫红色。分布于墨西哥的科阿韦拉州。

Flowers large, purplish-pink. Distributed in Coahuila State, Mexico.

长刺虾
Echinocereus longisetus (Engelm.) Lem.

植株长球形，从基部开始分枝或丛生，生有 10~14 根绒毛状白色长周刺，3~5 根白色长中刺，花大、侧生，深红色，喇叭状花心黄色，花蕊翠绿色。原产墨西哥和美国南部。

Whole plant long spherical, branched or clustered at base; flowers big, dark red, and lateral; flower centre trumpet shaped and yellow, stamen emerald; native to Mexico and southern U.S.

春高楼
Echinocereus palmeri Britton & Rose

植株小分枝，具 6~10 条棱，周刺 9~15 根，白色，尖部黑色，中刺 1~2 根，长达 2 cm，尖部颜色较深，长达 7 cm，紫丁香色，喉部较浅。原产美国新墨西哥州和墨西哥北部。

Whole plants branched , 6 to 10 ribs; 9 to 15 white outer spines with black apex, 1 to 2 central spines, up to 2 cm long; native to New Mexico, America or north Mexico.

三光球
Echinocereus pectinatus (Scheidw.) Engelm.

植株长球形至筒形，长有20~23条棱，周刺锯齿状，25~30根，中刺2~6根，全为白色、黄色、红色或深褐色。花紫红色，浅绿色花心。原产墨西哥。

Whole plant long spherical or terete; 20 to 23 ribs; 25 to 30 jagged outer spines, 2 to 6 central spines, all of them are white, yellow, red, or dark brown; flowers purple red; native to Mexico.

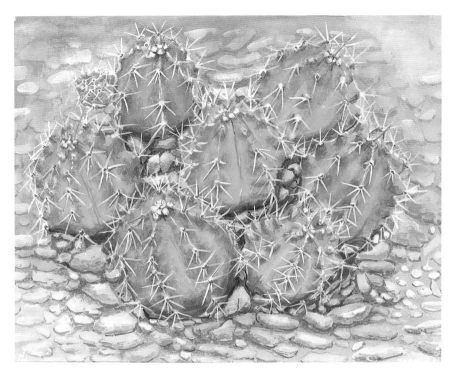

硬刺鹿角柱
Echinocereus rigidissimus (Engelm.) Haage

植株长筒形，20~22 条柱，8~10 根短而硬的周刺，棕红色，小分枝。花硕大，紫红色。原产美国南部。是一种颇受种植者喜爱的种类。

Whole plants terete; flowers big and purple-red; native to southern U.S.; deeply favored by plants enthusiasts.

顶花虾
Echinocereus pulchellus (Mart.) K. Schum.

植株单生或分枝，9~17 条棱，具疣突，周刺 13~15 根，紧贴茎体，无中刺，花粉红色或白色。分布于墨西哥的瓦哈卡州北部、普埃布拉州、伊达尔戈州和克雷塔罗州东南部。

Whole plant solitary or branched; 9 to 17 ribs; 13 to 15 outer spines are tightly attached to the stems, no central spines; flowers pink or white; distributed in northern Oaxaca, Puebla, Hidalgo and southeastern Queretaro States, Mexico.

太阳
Echinocereus rigidissimus subsp. rubispinus (G. Frank & A.B. Lau) N. P. Taylor

花大，粉红色。分布于墨西哥的索诺拉州的北部，美国新墨西哥州西南部和亚利桑那州东南部。

Flowers large, brilliant pink. Distributed in northern Sonora, Mexico and southwestern New Mexico, southeastern Arizona, the United States.

草市角
Echinocereus scheeri (Salm-Dyck.) Rumpler

植株匍匐生长，分枝，粗 2.5~3 cm，具 8~10 条略带突起的棱，周刺 7~12 根白色，底端黄色，中刺 1~3 根，褐色，花玫瑰红色。原产墨西哥。

Whole plant sprawling and branched; diameter 2.5 to 3 cm; 8 to 10 ribs; 7 to 12 white outer spines with yellow bottom, 1 to 3 brown central spines; flowers rose red; native to Mexico.

大花虾
Echinocereus sp. (Hybrid)

植株柱形，8~12 条棱，周刺 8~10 根，中刺 1 根，棕色，花淡红至白色，硕大，为栽培种。

A cultivar; whole plant terete, 8 to 12 ribs; 8 to 10 outer spines, 1 brown central spine; flowers pale red to white, big.

稀刺鹿角柱
Echinocereus subinermis Salm-Dyck ex Scheer

花大，黄色，花柄特别长，有刺。分布于墨西哥的锡那罗亚州北部、索诺拉州南部和奇瓦瓦州西南部。

Flowers large, yellow, with long flower stalk covered with spines. Distributed in northern Sinaloa, southern Sonora and southwestern Chihuahua States, Mexico.

黑花青花虾
Echinocereus triglochidiatus Engelm.

植株球形，基本不分枝，13~15 条棱，周刺 13~15 根，中刺 2~3 根，与青花虾不同，黑花青花虾花开亮红色，雌蕊绿色。原产美国。

Whole plant spherical, rarely branched; 13 to 15 ribs; 13 to 15 outer spines, 2 to 3 central spines; flowers bright red, pistil green; native to the United States.

微刺鹿角柱
Echinocereus subinermis var. ochoterenae (J.G. Ortega) G. Unger

植株生长初期为球形，后变长成筒形，多分枝，有 5~9 条棱，初期有 3~8 根周刺，1 根中刺，黄色。生长后期刺座长出 3~4 根 1 mm 刺。原产墨西哥。

Newborn plants spherical, gradually turn into terete; multi-branch; 5 to 9 ribs; newborns have 3 to 8 spines, 1 yellow central spine, mature ones grow 3 to 4 spines from the areole; native to Mexico.

美花鹿角柱
Echinocereus triglochidiatus Engelm.

花大，粉色或白色。分布于美国亚利桑那州东南部、新墨西哥州南部和西部、得克萨斯州西部和墨西哥的奇瓦瓦州北部。

Flowers large, pink or white. Distributed in southeastern Arizona, southern and western New Mexico, western Texas, the United States and northern Chihuahua, Mexico.

篝火
Echinocereus triglochidiatus var. melanaeanthus (Engelm.) L.D. Benson

植株圆柱形，多分枝，生有6~10棱，周刺3~5根，中刺1~5根，花猩红色，长喇叭型。原产美国。

Whole plants terete, multi-branched; 6 to 10 ribs; 3 to 5 outer spines, 1 to 5 central spines; flowers scarlet, long trumpet shaped; native to America.

美花鹿角柱变种
Echinocereus triglochidiatus var. paucispinus Engelm. ex W.T. Marshall

花大，橙红色。分布于美国南部和墨西哥。
Flowers orange red. Distributed in southern of the United States and Mexico.

三刺虾
Echinocereus triglochiidiatus var. melanacanthus (Engelm.) L. D. Benson/ Wilcoxia schmollii

多分枝，群生，碧绿色，圆锥形球体，具 8 条棱。刺座着生于棱缘，每一刺座着生 3 枚白色周刺。花大，深红色，漏斗状。分布于美国新墨西哥州、亚利桑那州东北部、加利福尼亚州南部和德克萨斯州南部。

Usually clustered; whole plants green, conical spherical; 8 ribs, areoles grow on the edges of these ribs, 3 white outer spines grow on every areole; flowers big and deep red, funnel like; distribute in New Mexico, northeastern Arizona, southern California and Texas, the United States.

青花虾
Echinocereus viridiflorus Engelm.

植株球形，极少分枝，粗 2~5 cm，具 13~15 条直棱，刺座略长，周刺 13~15 根，中刺 2~3 根或更少，刺较短，白色或褐色。花绿色，外瓣具较深色中分条纹。原产美国南部。

Whole plant spherical, rarely branched, diameter 2 to 5 cm; 13 to 15 straight ribs; 13 to 15 outer spines, less than 3 central spines, all of them short, white or brown; areole relatively long; flowers green, outside petals with dark colored stripes; native to Southern U.S..

威氏鹿角柱
Echinocereus websterianus G.E. Linds.

植株筒形，由 20~22 条棱构成，花靠顶生，粉红至红色，花心淡绿。原产美国西南（部）和墨西哥。

Whole plant terete; 20 to 22 ribs; flowers acrogenous, petals pink to red, centre light green; native to southwestern U.S. and Mexico.

珠毛柱
Echinocereus schmollii (Weing.) N. P. Taylor [*Wilcoxia schmollii* (Weing.) F.M. Knuth]

为灌木形仙人掌，株高 25 cm，茎粗 1 cm，嫁接后茎可达 2 cm，具 8~10 条棱，棱上具疣突，周刺细，35 根，花多为浅粉色。原产墨西哥。

Whole plant shrub like, 25 cm tall, stem diameter 1 cm; 8 to 10 ribs with warts; 35 thin spines; flowers pale pink; native to Mexico.

雪溪
Echinofossulocacus albatus A. Dietr.

植株单生，扁球形或球形。球径可达 12 cm，深灰绿色；具多达 30~36 条波浪状薄棱；具辐射状淡黄色周刺 10~12 枚和 4 枚较粗较长中刺，包裹整个球体；花白色，漏斗状，顶生；花期初春。原产墨西哥中部高原地区。

Whole plants solitary, flat spherical or spherical, diameter 12 cm, dark greyish green; 30 to 36 ribs; 10 to 12 yellow outer spines and 4 thick, long central spines; acrogenous flowers are white and funnel shaped; florescence is in early spring; native to central Mexico.

多棱玉
Echinofossulocactus multicostatus (Hildm.) Britton & Rose

植株多为单生，球形或扁球形，球径约 10 cm，碧绿色。具 80~100 条薄纸板状棱。刺座长有白色绒毛，着生中刺 3 枚，长短不一。花顶生，白色，有粉紫红色条纹，有时会出现黄色纹；初春开花。原产墨西哥中部高原。

Whole plant usually solitary, spherical or flat spherical, diameter 10 cm, aquamarine; 80 to 100 ribs; white tomentum grow on areole; flowers white, and acrogenous with pink-purple or yellow stripes; florescence is in early spring; native to central Mexico.

剑恋玉锦
Echinofossulocactus kellerianus 'Variegata'

为剑恋玉的一个黄斑栽培变种。
A cultivar of Echinofossulocactus cactus kellerianus with yellow spots.

千波万波锦
Echinofossulocactus multicostatus 'Variegata'

为千波万波的一个黄斑栽培变种

A cultivar of Echinofossulocactus multcostus with yellow spots.

五刺玉
Echinofossulocactus pentacanthus (Lem.) Britton & Rose

植株扁平球形，绿色，24~28条波浪式薄棱，周刺5根，植株顶刺密生，棕色，长约8 cm，花浅红色。原产美国南部，墨西哥北部。

Whole plant flat spherical, green; 24 to 28 wave like ribs; 5 outer spines, brown, 8 cm long spines grown on top; flowers pale red; native to southern U.S. and northern Mexico.

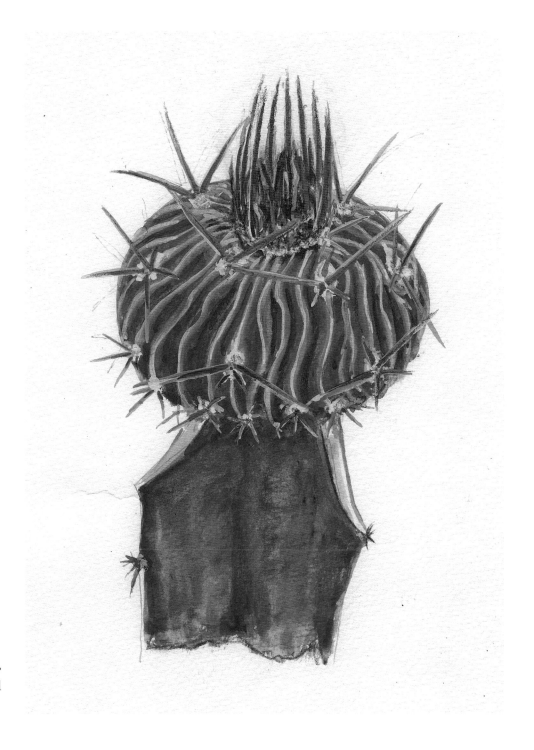

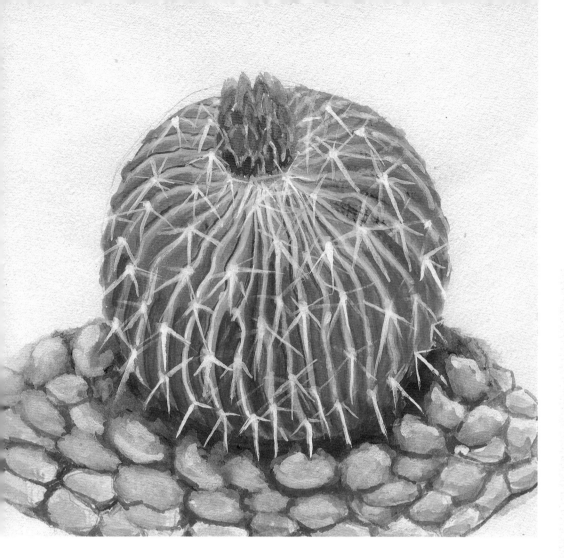

太刀岚
Echinofossulocactus phyllacanthus (Mart.) Lawr.

植株初生为单生，成长后会萌生出子球，呈群生。球形至扁球形。具30~35条波浪形薄棱。周刺2~7枚，上端3根，扁平、螺旋状或环形。新刺红色或酒石红色，老刺为褐色，长4~8 cm。花黄白色，具红色棕条纹。原产墨西哥中部高原。

Newborn plants solitary, clustered after mature; spherical to flat spherical; 30 to 35 wave like ribs; 2 to 7 flat, spiral or circular outer spines, colors ranging from red to brown, 4 to 8 cm long; flowers yellowish white with red-brown stripes; native to central Mexico.

龙舌玉（秋阵管）
Echinofossulocactus vaupelianus (Werderm.) Oehme

植株单生，扁球形至球形，暗绿色，球径约8 cm。具30~40条波浪形纵向薄棱，具针状周刺10~25枚和中刺1~4枚，非常坚硬。顶部中刺向上伸展，呈黄色、褐色或黑色，长约7 cm。花呈白色、黄色或粉红色。原产墨西哥。

Whole plant solitary, flat spherical to spherical, dark green, diameter 8 cm; 30 to 40 wave like thin ribs; 10 to 25 needle like outer spines, 1 to 4 hard central spines; flowers white, yellow or pink; native to Mexico.

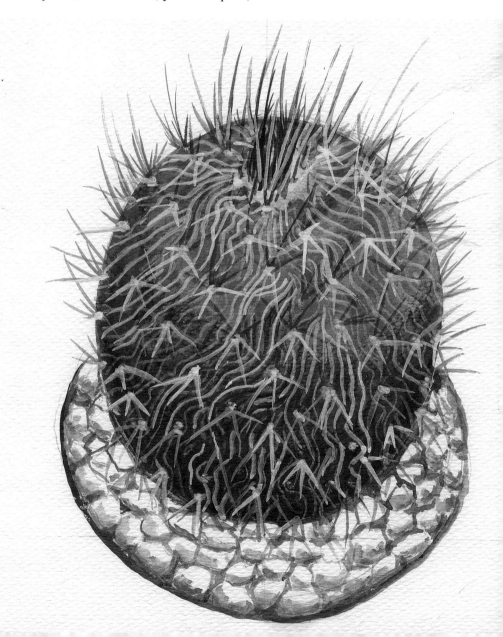

缩玉
Echinofossulocactus zacatecasensis Britton & Rose

植株单生，扁平球形至球形，绿色至灰绿色，球径约 10 cm。36~55 条褶皱弯曲的棱，具 10~12 枚长 1 cm 的周刺和 3 枚长 4 cm 的中刺，靠球顶部中刺向上伸展。花淡紫色，具深紫红色纵向条纹。原产墨西哥中部高原。

Whole plant solitary, flat spherical or spherical, green to greyish green, diameter 10 cm; 36 to 55 ribs; 10 to 12 outer spines and 3 central spines; flower pale purple, with dark purple-red stripe; native to central Mexico.

缩玉锦
Echinofossulocactus zacatecasensis 'Variegata'

为缩玉的一个黄斑栽培变种。
A cultivar of Echinofossulocactus zacatecsenis with yellow spots.

仙人球 – 杂交栽培种
Echinopsis (Hybrid)

金城
Echinopsis candicans (Gillies in Salm-Dyck.) D.R. Hunt (Hybrid)

植株直立或匍匐生长, 株高1米, 株幅10cm, 草绿色或蓝绿色, 具9~11条棱, 周刺11~14根, 长4cm, 中刺1~4根, 长10cm, 花白色, 夜间开放, 原产阿根廷。

Whole plant erect or procumbent, up to 1 m high, 10 cm wide, green or blueish green; 9 to 11 ribs; 11 to 14 outer spines, up to 4 cm long, 1 to 4 central spines, about 10 cm long; flowers white, bloom at night; native to Argentina.

沙漠柱
Echinopsis deserticola (Werderm.) Friedrich & G.D. Rowley

植株柱状, 高1~1.5米, 株幅4~7 cm, 草绿色或灰绿色, 带8~13条深深的凹陷的棱, 周刺9~12根, 中刺2~4根, 初为深褐色, 后变灰色, 花侧生, 白色, 夜间开放, 有香味。原产智利。

Whole plant terete; 1 to 1.5 m tall, 4 to 7 cm wide, green or greyish green; 8 to 13 ribs; 9 to 12 outer spines and 2 to 4 central spines; flowers white, fragrant and nocturnal; native to Chile.

年氏仙人球
Echinopsis formosa (Pfeiff.) Jacobi ex Salm-Dyck.

植株球形或长球形，长有19~20条浅棱，周刺8~12根，白色，中刺2~4根，棕黑色，较坚硬。花顶生，深红色。原产南美洲玻利维亚和阿根廷。

Whole plant spherical or long spherical; 19 to 20 ribs; 8 to 12 white outer spines, 2 to 4 hard, brown central spines; flowers dark red, grow on top; native to Bolivia and Argentina.

倒波形仙人球
Echinopsis obrepanda (Salm-Dyck) K. Schum.

花为红色或洋红色，花期在5~6月。分布于玻利维亚。

Flowers red or magenta, florescence is from May to June. Distributed in Bolivia.

旺盛球
Echinopsis oxygona (Link) Zucc.

植株直径最长可达 25 cm，周刺 5~15 根，中刺 0~5 根，花粉红色，夜间开放。原产阿根廷北部。

Diameter up to 25 cm; 5 to 15 outer spines and 0 to 5 central spines; flowers pink and nocturnal; native to the north of Argentina.

红花毛花柱
Echinopsis smrziana Backeb.

植株长球形，深绿色，具 12~16 较深的棱，周刺 8~12 根，白色，中刺 1~2 根，褐色或黑色，花较大，有白色，黄色，也有红色。原产南美洲。

Whole plants long spherical, dark green; 12 to 16 ribs; 8 to 12 white outer spines, 1 to 2 brown or black central spines; flowers big, white, yellow or red; native to South America.

圆齿昙花
Epiphyllum crenatum (Lindl.) G. Don

植株与昙花相似，生有叶片状扁平无刺的茎，灰绿色，花柄较长，外瓣黄色，内瓣白色。原产墨西哥和危地马拉。

Whole plant looks like E. oxypetalum; stems flat, leaf like and spineless; outside petals yellow, inside petals white; native to Mexico and Guatemala.

多花昙花
Epiphyllum floribundum Kimnach

植株生有叶片状扁平无刺的茎，灰绿色，花柄绿色，外瓣棕红色，内瓣浅黄色和白色。原产墨西哥等。

Stems flat, leaf like and spineless, greyish green; outside petals red-brown, inside petals pale yellow and white; native to Mexico.

昙花
Epiphyllum oxypetalum (DC.) Haw.

为叶花类仙人掌，茎上有深棱沟，直径长达12 cm，花大，外瓣红色，内瓣白色。原产墨西哥，危地马拉，委内瑞拉和巴西。

Diameter up to 12 cm; flowers big, outside petals red, inside petals white; native to Mexico, Guatemala, Venezuela and Brazil.

昙花花蕾
Epiphyllum oxypetalum (DC.) Haw. *

月世界
Epithelantha micromeris (Engelm.) F.A.C. Weber ex Britton & Rose

植株单生，短小球形，直径约 2~2.5 cm，具 30~38 条由小疣突排列成螺旋状的棱，刺座长于疣突尖上，具 18~26 枚白色针状紧贴棱的短刚刺和 1~2 枚白色较长针刺。花顶生，粉红色。果实红色。不易栽种，但为爱好者非常钟爱的小型品种。原产美国德克萨斯州西部和墨西哥北部。

Whole plant solitary, small and spherical; 30 to 38 spiral ribs covered with little warts; areole grow on wart apex, 18 to 26 white, short, needle like spines tightly attached to the ribs; flowers pink, grow on top; not easy to cultivate; deeply favored by plants enthusiasts; native to Texas, U.S. and northern Mexico.

小松球
Escobaria minima (Baird) D. R. Hunt

植株长圆形，具 2 mm 长疣突，周刺 13~18 根，呈辐射状，蜂蜜黄或更深一点，长 1 cm，无中刺，花粉紫红色。原产美国，已被列入珍稀美丽保护物种。

Whole plant long spherical; with 2 mm long warts; 13 to 18 yellow outer spines, up to 1 cm long, no central spines; flowers purple red; native to America.

天龙
Escobaria strobiliformis (Poselger) Scheer ex Boed.

植株圆柱形，株高 18 cm，株幅 6 cm。周刺 20~30 根，白色，长 0.5~1.5 cm，中刺 5~9 根，白色，刺尖黑色，坚硬且较长；花粉红，外瓣为紫红色。原产美国和墨西哥北部。

Whole plants terete, up to 18 cm high, 6 cm wide; 20 to 30 white, 0.5 to 1.5 cm long outer spines, 5 to 9 white, long, and hard central spines with black apex;; flowers pink, outside petals purple red; native to U.S. and north Mexico.

秘鲁毛柱
Espostoa huanucoensis Johnson ex F. Ritter

植株丛生，圆柱状，生有 12~16 条浅直棱，刺座长有白色绒毛，植株顶长有密生的绒毛。原产秘鲁。

Whole plant clustered, terete; 12 to 16 straight ribs; white tomentum grow on areole; native to Peru.

老乐柱
Espostoa lanata (Kunth) Britton & Rose

　　植株生长成熟后会分枝，呈烛台形或树形，高可达 4 米，茎粗约 15cm，具 20~30 条浅棱，周刺约 12 枚，中刺 1~2 枚，最长达 8cm，刺座着生许多棉状绒毛，白色或浅黄金色，包裹整个植株，花白色侧生，分布于秘鲁、南美洲等地区，现已广泛引种到世界各地。

　　Whole plant tree like or candlestick shaped, branched after mature, up to 4 m high, stem diameter 15 cm, 20 to 30 shallow ribs; 12 outer spines, 1 to 2 central spines, up to 8 cm long;; white or pale golden tomentum grow on areole; flowers white and lateral; distributed in Peru and South America, widely introduced all around the world.

小老乐
Espostoa lanata (Kunth) Britton & Rose (Small)

花为白色。分布于秘鲁。植株体形较矮小。
Flowers white. Distributed in Peru.

短毛老乐
Espostoa lanata (Kunth) Britton & Rose

为老乐一个短毛刺变种，植株丛生，圆柱形，生有16~20条浅直棱，绒毛状刺很短，顶部长较多绒毛状周刺。原产秘鲁。

A short and spineless variant of E. lanate; whole plants clustered, terete; 16 to 20 shallow, straight ribs; spines hair like and short; native to Peru.

老乐花
Espostoa lanata (Kunth) Britton & Rose (Flowers)

短毛花壶柱
Eulychnia breviflora Phil.

植株灌木形，茎高达 3 米，粗 6~10cm，具 10~13 条棱，周刺 10~22 根，最长 3cm，中刺 3~6 根，最长 15cm，花浅粉红色或浅黄至白色，原产智利。

Whole plant shrub like, stem up to 3 m, diameter 6 to 10 cm, 10 to 13 ribs; 10 to 22 outer spines, up to 3 cm long, 3 to 6 central spines, up to 15 cm long; flowers pale pink, pale yellow or white; native to Chile.

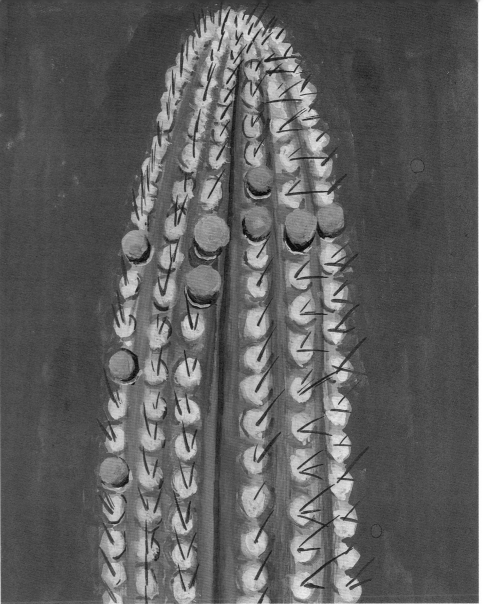

白幻阁
Eulychnia ritteri 'Cullman'

植株树形或灌木形，很多分枝，茎高可达 4 米，茎粗 6~8 cm，具 13~20 条棱，周刺约 12 根，褐色，中刺 1~4 根，黑色，长 3~8 cm，花粉红色。原产秘鲁。

Tree or shrub like, branched; stem up to 4 m high, diameter 6 to 8 cm; 13 to 20 ribs; 12 brown outer spines, 1 to 4 black, 3 to 8 cm long central spines; flowers pink; native to Peru.

彩色强刺球
Ferocactus coloratus H.E. Gates

花为稻草色。分布于墨西哥下加利福尼亚州。

Flowers straw-colored. Distributed in Baja California, Mexico.

琥头
Ferocactus cylindraceus (Engelm.) Orcutt

花为黄色或绿色。分布于美国加利福尼亚州南部、内华达州、亚利桑那州和墨西哥索诺拉州。

Flowers yellow or green. Distributed in southern California, Nevada, Arizona, the United States and Sonora, Mexico.

埃氏强刺球（江守）
Ferocactus emoryi (Engelm.) Orcutt

花为红色或黄色。分布于美国亚利桑那州的西南部和墨西哥的索诺拉州到锡那罗亚州的北部。

Flowers red or yellow. Distributed in southwestern Arizona, the United States and Sonora to northern Sinaloa, Mexico.

刈穗玉
Ferocactus gracilis H.E. Gates

花为红色。分布于墨西哥下加利福尼亚州中北部。

Flowers red. Distributed in central northern Baja California, Mexico.

红刺强刺球

Ferocactus hamatacanthus (Muehlenpf) Britton & Rose

花为紫色、粉色。分布于墨西哥的普埃布拉、韦拉克鲁斯等地。

Flowers purple or pink. Distributed in Puebla and Veracruz, Mexico.

大虹

Ferocactus hamatacanthus (Muehlenpf) Britton & Rose

单生，球形至长圆筒形，高 40~50 cm，茎粗 13~15 cm，暗绿色。具 12~16 条疣突起脊高的棱，着生黄白色至灰白色针状钩刺 1~3 枚。花顶生，漏斗状，黄色。

Solitary, spherical or long cylindrical; 40 to 50 cm high, stem diameter 13 to 15 cm, dark green; 12 to 16 verrucose ribs covered with 1 to 3 hooked, needle like spines, colors ranging from yellowish white to greyish white; funnel like flowers are acrogenous and yellow.

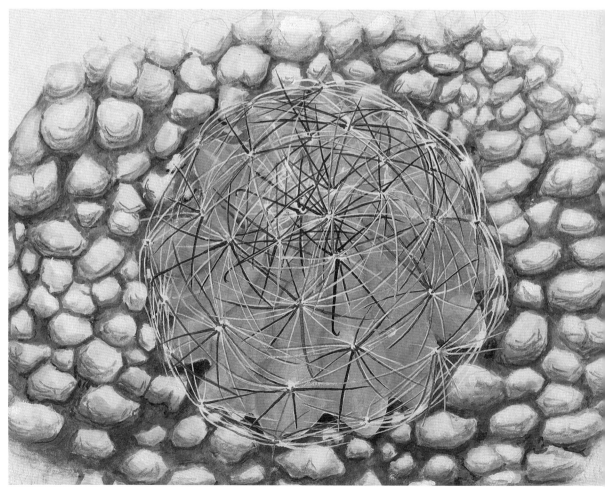

恰氏强刺球
Ferocactus herrerae J.G. Ortega

植株一般单生，椭圆形至圆筒形，绿色。具14~16条纵向具节的直棱，刺座着生于棱节上，具8~10枚白色细短周刺和5~6枚棕黑色较粗壮的钩刺，花黄色，顶开生。原产美国亚利桑那州和墨西哥北部。

Usually solitary, oval to terete, green; 14 to 16 straight ribs; 8 to 10 white, tiny outer spines, 5 to 6 brown, thick, hooked spines; flowers acrogenous and yellow; native to Arizona, U.S. and northern Mexico.

箭状强刺球（变种）
Ferocactus histrix (DC.) G. E. Linds

日之出（日出）
Ferocactus latispinus (Haw.) Britton & Rose

植株一般单生。具 12~16 条具疣突的棱。刺座长在疣突的棱缘上，具 12~16 枚白色毛发状周刺和 4~6 枚淡红色针状周刺，2 枚红色扁平长约 2~3 cm 的钩刺。靠植株顶部的红色周刺和中刺都往上伸展。花顶生，红色。花蕾具枣红色纵向条纹。原产墨西哥中部和西部山区。

Whole plant usually solitary; 12 to 16 verrucose ribs; 12 to 16 white, hair like and 4 to 6 pale red, needle like outer spines, 2 flat, red hooked spines; flowers acrogenous and red; flower buds have purplish red stripes; natvie to central and western Mexico.

箭状强刺球
Ferocactus histrix (DC.) G.E. Linds

花为黄色，花期是 6~7 月。广布于墨西哥中部，圣路易斯波托西附近也有分布。

Flowers yellow, florescence is from June to July; Widespread in central Mexico, common in some places near San Luis Potosi city, Mexico.

林氏强刺球
Ferocactus lindsayi Bravo.

花为黄色。分布于墨西哥里约巴尔萨斯盆地的米却肯州和阿帕辛甘的东南部。

Flowers yellow. Distributed in Michoacan, of Rio Balsas basin, southeastern of Apatzingan, Mexico.

赤凤
Ferocactus pilosus (Galeotti) Werderm.

植株一般单生，椭圆形至筒形，碧绿色至灰绿色，具18~20条纵向薄直棱。棱缘脊上着生较密的刺座，具10~14枚放射状向内伸展的周刺和1~2枚暗红色粗壮具钩的中刺。顶部的钩刺直向上伸展。花钟状，顶生，黄色。原产墨西哥和美国德克萨斯州，现被引种到各地栽培。

Whole plant usually solitary, oval or cylindrical, aquamarine to greyish green; 18 to 20 thin straight ribs; 10 to 14 ridial outer spines extend inward, 1 to 2 thick, hooked, dard red central spines; dense areoles grow on the margin of ribs; flowers campanulate, yellow; native to Mexico and Texas, U.S.; widely introduced nowadays.

波氏强刺球
Ferocactus pottsii (Salm-Dyck.) Backeb.

花为黄色。分布于墨西哥奇瓦瓦州西南部、索诺拉州东南部和锡那罗亚州北部。

Flowers yellow. Distributed in southwestern Chihuahua, southeastern Sonora and northern Sinaloa, Mexico.

波氏强刺球
Ferocactus pottsii (Salm-Dyck.) Backeb.

植株高大，株幅达 50 cm，高 1 m，具 12~16 条肥大棱，刺座不多，每条棱缘上着生 5~6 个刺座，周刺 6~8 根，中刺 2~3 根，棕黑色。原产墨西哥等地。

Whole plant; up to 50 cm wide, 1 m tall; 12 to 16 fleshy ribs; 5 to 6 areoles grow on the rib margin; 6 to 8 outer spines, 2 to 3 central spines; native to Mexico.

长刺强刺球
Ferocactus rectispinus 'Longispinus'

长刺荒鹫
Ferocactus reppenhagii 'Longispinus'

植株呈扁球型，株幅约 80 cm，高 24 cm，周刺 6~8 根，中刺 1~3 根，黄色，长约 6~8 cm，坚硬，花顶生，金黄色。原产美国加利福尼亚州和墨西哥。

Whole plant flat spherical, 80 cm wide, 24 cm tall; 6 to 8 outer spines, 1 to 3 yellow central spines, 6 to 8 cm long; flowers acrogenous and golden; native to California, U.S. and Mexico.

黄彩玉
Ferocactus schwarzii Linds

为较大型强刺球，植株扁平球形至球形，14~16 条较深的具横肋的棱，刺座密生于棱缘上，刺稀疏，针状白色。花钟状，顶生，黄色。原产墨西哥北部和美国西南部。

Whole plant flat spherical to spherical; 14 to 16 ribs; areoles grow on arris margin; flowers campanulate, yellow, grow on top; native to northern Mexico and southwestern United States.

勇状球
Ferocactus robustus (Link & Otto) Britton & Rose

植株群生，单柱可达 20cm，生有 8 条棱，周刺 10~14 枚，中刺 6~12 枚，最长可达 6cm，花黄色，原产墨西哥。

Whole plants clustered, 8 rib; 10 to 14 outer spines, 6 to 12 central spines, up to 6 cm long; flowers yellow; native to Mexico.

短刺强刺球
Ferocactus sp.

植株扁平球形，长有 8 条三角形宽棱，灰蓝绿色，每条棱上长有 2 刺座，短周刺 8~10 根。原产不详。

Whole plant flat spherical; 8 triangular ,broad, greyish blue-green ribs; 2 areoles grow on every ribs, 8 to 10 short outer spines.

钩刺强刺球（变种）
Ferocactus townsendianus Britton & Rose

植株球形，具 16~18 条直棱，生有 8~12 根灰白色周刺，1~2 根较长粗周刺，刺突具钩，黑色。原产墨西哥。

Whole plant spherical, 16 to 18 straight arrises; native to Mexico.

钩刺强刺球
Ferocactus townsendianus Britton & Rose

植株倒卵形，长有 12~14 条直棱，周刺 12~16 根，白色，约 1~2 cm 长，中刺，1~2 根，棕红色，坚硬钩刺。原产墨西哥。

Whole plant obovate, 12 to 14 straight ribs, 12 to 16 white, 1 to 2 cm long outer spines, 1 to 2 brownish red, hard, hooked central spines; native to Mexico.

赤金龙（变种）
Ferocactus wislizeni (Engelm.) Britton & Rose

赤金龙（野外生长）
Ferocactus wislizeni (Engelm.) Britton

赤金龙（温室栽种）
Ferocactus wislizeni (Engelm.) Britton

植株单生，圆柱形，为强刺属高大种类，高可达1.5 m以上，茎粗80 cm以上，具18~28条直棱。大部分植株灰绿色，少部分深绿色。刺座着生于棱缘上，密生，具10~12枚白色周刺。1~2枚中刺。花钟状，顶生，黄色，有时也见红色和红褐色的花，群生于植株顶部。花期夏季，果实黄色。原产美国亚利桑那州和墨西哥北部。

Whole plant solitary, terete, up to 1.5 m tall, stem diatmeter more than 80 cm, mostly greyish green, sometimes dark green 18 to 28 straight ribs, areoles grow on rib margin; 10 to 12 white outer spines, 1 to 2 central spines; acrogenous flowers are campanulate, yellow, red, or brownish red; florescence is in summer, fruits are yellow; native to Arizona, U.S. and northern Mexico.

士童
Frailea castanea Backeb.

花为黄色。分布于巴西南部到乌拉圭北部。

Flowers yellow. Distributed in southern Brazil to northern Uruguay.

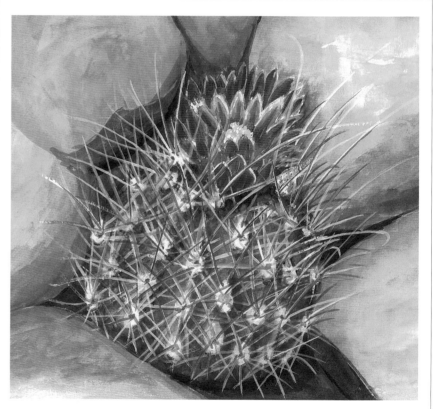

钩刺球
Glanodulicactus uncinatus (Galeotti ex Pfeiff.) Backeb.

植株球形，具12~14条直棱，6~10根灰白色周刺，1~3根带钩的中刺，花顶生，红色。原产美国西南部和墨西哥。

Whole plant spherical, 12 to 14 straight arrises; flowers red and apical; native to southwestern U.S. and Mexico.

次大陆裸萼球
Gymnocactus subterraneus (Backeb.) Fritz Schwarz

花为洋红色。分布于墨西哥新莱昂州和阿兰贝里。

Flowers magenta. Distributed in Nuevo Leon and Aramberri, Mexico.

翠晃冠（翠花冠）
Gymnocalycium anisitsii (K. Schum.) Brltton et Rose

植株球形，鲜绿色或橄榄绿色，不易分枝。株幅宽 7~10 cm，刺有 5~7 根，具 6~8 条宽阔直棱，周刺 6~8 根，白色。花外瓣淡紫色，内瓣白色。原产巴西；现已被广泛引种。

Whole plant spherical, bright green or olive green; not easy to branch; 7 to 10 cm wide; 5 to 7 spines; 6 to 8 broad, straight ribs; outside petals pale purple, inside petals white; native to Brazil, widely introduced.

绯花玉
Gymnocalycium baldianum (Speg.) Speg.

植株球形，深灰色或蓝绿色，株幅宽 7~10cm，具 12~14 条浅直棱，周刺 5~7 根，花浅红色或深红色。原产阿根廷，生长较粗犷。

Whole plant spherical, dark grey or blue-green, 7 cm wide; 12 to 14 shallow, straigh ribs; 5 to 7 outer spines; flowers pale red or dark red; native to Argentina, easy to cultivate.

罗星球
Gymnocalycium bruchii (Speg.) Hosseus

植株深绿色，分枝，株幅 3~6 cm，周刺 10~12 根，中刺 0~3 根，刺短，有的紧贴植株。花淡粉红色，花蕾紫红色，顶生。原产阿根廷。

Whole plants dark green, branched, 3 to 6 cm wide; 10 to 12 outer spines, 0 to 3 short central spines; acrogenous flowers are pale pink, flower buds purple-red; grow on top; native to Argentina.

圣王丸
Gymnocalycium buenekeri (Buining) Swales

花为粉红色。分布于巴西南里奥格兰德州。
Flowers pink. Distributed in Rio Grande do Sul, Brazil.

卡迪那斯裸萼球
Gymnocalycium cardenasianum F. Ritter

花为粉色到白色，分布于玻利维亚南部。
Flowers pink to white. Distributed in southern Bolivia.

天王球
Gymmocalycium dnudatum (Link & Otto) Pfeiff. ex Mittler

植株亮深绿色，球形，具 5~8 扁平棱，周刺 5~8 根，幼小紧贴植株，花酒红色有光泽。原产阿根廷东北部，巴西南部，乌拉圭北部。
Whole plants bright green, spherical, 5 to 8 flat rbis; 5 to 8 tiny outer spines; flowers wine red, glossy; native to northeastern Argentina, southern Brazil, and northern Uruguay.

胭脂牡丹锦

Gymnocalycium mibanovichii var. friedrichii 'Pink Botan Niahiki Variegata'

米氏裸萼球

Gymnocalycium mihanovichii (Frič & Gürke) Britton Rose

花为黄白色到粉红色。分布于巴拉圭的查科省。

Flowers yellowish-white to pink. Distributed in Chaco, Paraguay.

黑牡丹玉
Gymnocalycium mihanovichii var. friedrichii 'Black'

植株墨绿至灰黑，生有 8 条阔尖棱，株幅宽约 8 cm，有的边缘有横边，周刺 4~6 根，花粉红色。原产巴拉圭。

Whole plant dark green to greyish black; 8 cm wide; 8 broad, sharp ribs; 4 to 6 outer spines; flowers pink; native to Paraguay.

绯牡丹彩色裸萼球
Gymnocalycium mihanovichii var. friedrichii 'Hibotaii Nishiki'

植株红色并带翠绿色的斑昙，生有 8 条阔尖棱，株幅宽约 8 cm，边缘有横边，周刺约 6 根，花外瓣淡绿色，内瓣淡粉红色。原产巴拉圭。

Whole plant red; 8 broad, sharp ribs, 8 cm wide; outside petals pale green, inside petals pale pink; native to Paraguay.

凤牡丹
Gymnocalycium mihanovichii var. friedrichii 'Koki Variegata'

鸡冠绯牡丹
Gymnocalycium mihanovichii var. friedrichii 'Rubra-crista'

该种为绯牡丹缀化变种，植株的生长点呈弯曲扇形，使植株呈鸡冠状，红色。植株由于缺少叶绿素必须嫁接种植。该种形状特别，颇受爱好者喜爱。原产南美洲。
A variant of G. mihanovichii; whole plants cristate, red; native to South America.

珠红绯牡丹
Gymnocalycium mihanovichii var. friedrichii Werderm

五彩绯牡丹
Gymnocalycium mihanovichii var. friedrichii Werderm.

祥云锦
Gymnocalycium mihanovichii var. stenogonum 'Variegata'

植株扁球形，生有8条阔尖棱，边缘有横边，周刺4~6根，花内瓣粉红色，外瓣有时绿色。原产巴拉圭。

Whole plant flat spherical, 8 broad, sharp ribs; 4 to 6 outer spines; inside petals pink, outside petals sometimes green; native to Paraguay.

瑞云球（金色）
Gymnocalycium mihanovichii var. Stenosonun 'Variegata' var. friedrichii 'Aurea'

植株灰绿色或紫褐色，也有金黄色。生有8条阔尖棱，株幅宽8 cm，有的边缘有横边。小棱宽可达6 cm，周刺5~6根，花瓣外瓣绿色，内瓣为白色或粉红色。原产巴拉圭。

Whole plant greyish green or purple-brown, sometimes golden yellow; 8 cm wide; 8 broad, sharp ribs; 5 to 6 outer spines; outside petals green, inside petals white or pink; native to Paraguay.

牡丹玉
Gymnocalycium mihanovichii (Frič & Gürke) Britton & Rose

植株扁球形，灰绿色或紫褐色，小型品种，球径约 4 cm，生有 7~9 条具明显横沟的宽棱，尖棱宽可达 6 cm。刺座常见有白色绒毛，周刺 5~6 根，但很短。花着生于球顶，外花瓣绿色、红色，花白色或粉红色。不但是一种十分受欢迎的小型盆栽仙人掌植物，而且具有很多的变种。原产巴拉圭。

Whole plant flat spherical, greyish green or purple-brown; a small species, diameter 4 cm; 7 to 9 broad ribs; white tomentum grow on areole; 5 to 6 short outer spines; flowers acrogenous, outside petals green or red; native to Paraguay.

黑网孔裸萼球
Gymnocalycium nigriareolatum Backeb.

花为闪光的瓷白色。分布于阿根廷的卡塔马卡省。

Flowers shimmering porcelain-white. Distributed in Catamarca, Argentina.

龙头
Gymnocalycium quehlianum (F.Haage ex Quehl) Vaupel ex Hosseus

植株暗灰色，多带些红褐色，周刺5根，仅长0.5~1 cm，花顶生，白色，花心红色。原产阿根廷。

Whole plant dark grey; flowers acrogenous and white; native to Argentina.

新天地
Gymnocalycium saglionis (Cels) Britton & Rose

新天地
Gymnocalycium saglionis (Cels) Britton & Rose

植株一般单生，亦常见在球体萌生出子球成丛生。扁平较大型圆球，蓝绿色，具14~30条疣突棱，刺座着生于疣突顶部，7~15枚弯曲的白色周刺和1~3枚中刺。花侧生，白色或淡粉红色。果实红色。原产阿根廷北部和玻利维亚南部。

Usually solitary; flat spherical, bule-green; 14 to 30 verrucose ribs, areole grow on top of the warts; 7 to 15 white, curved outer spines, 1 to 3 central spines; flowers pleurogenous, white or pale pink; fruits are red; native to northern Argentina and southern Bolivia.

守殿玉
Gymnocalycium stellatum Speg.

植株灰绿褐色，株幅达 10 cm，周刺 3~5 根，紧贴植株生长，花白色，花心粉红色。原产阿根廷。

Whole plant greyish green, 10 cm wide; flowers white; native to Argentina.

沙氏裸萼球
Gymnocalycium saglionis (Cels) Britton & Rose

花为白色混杂粉色。分布于阿根廷的萨尔塔省、卡塔马卡省和图库曼省。

Flowers white tinged with pink. Distributed in Salta, Catamarca and Tucuman, Argentina.

分生裸萼球
Gymnocalycium tillianum Rausch

花为红色，花期在 5~6 月。分布于阿根廷的卡塔马卡、塞拉利昂安巴托。

Flowers red, florescence is form May to June. Distributed in Catamarca and Sierra Leone Ambato, Argentina.

乌拉圭裸萼球
Gymnocalycium uruguayense (Arechav.) Britton & Rose

植株扁平的球形，花为黄色，花期在 6 月。广布于乌拉圭和邻近的巴西南部。

Whole plant flat spherical; flowers yellow, florescence is in June; Widespread in Uruguay and neighboring southern Brazil.

威迪斯裸萼球
Gymnocalycium vatheri Buining

植株球形，橄榄绿色，株幅 4 cm，高 9 cm，具 12~14 条直棱，刺座长有白色棉毛。周刺 1~3 条，约 5 cm，贴紧球体。花顶生，白色。原产阿根廷。

Whole plant spherical, olive green, 4 cm wide, 9 cm tall; 12 to 14 ribs; 1 to 3 outer spines, about 5 cm long; white tomentum grow on areole; flowers white, grow on top; native to Argentina.

金焰柱
Haageocereus multangularis (Haw.) F. Ritter

株高达 1.5 米，茎粗 6cm，刺黄色或红色并带有白色刚毛，花橙粉红色，原产秘鲁。

Whole plant up to 1.5 m high, stem diameter 6 cm; spines yellow or red, covered with white bristle; flowers orange pink; native to Peru.

假昙花杂交种
Hatiora (Hybrid)

假昙花杂交种
Hatiora (Hybrid)

"巨犬" 假昙花

Hatiora 'Grande'

假昙花

Hatiora epiphylloides (Porto & Werderm.) P.V.Heath

植株为叶状仙人掌，2~2.5 cm 宽，4~7 cm 长，刺座有 1~3 根刚毛，花为猩红色，4~5 cm 长，花萼长而尖。原产巴西。

Whole plant leaflike, 2 to 2.5 cm wide, 4 to 7 cm long; 1 to 3 bristle like spines grow on areole; flowers scarlet; native to Brazil.

特美牡丹柱
Heliocereus speciosus (Cav.) Britton & Rose

植株多直立或悬挂附生，茎初生为红绿色，后渐变为深绿色，长约1米，粗2~3 cm，具3~5条棱，刺5~8根，随生长刺会增加，刺黄色或褐色。花大，鲜紫红色，属仙人掌最美的花。原产于墨西哥中部。

Whole plant erect or epiphytic; newborn stems red-green, gradually turn into dark green, up to 1 m high, diameter 2 to 3 cm; 3 to 5 ribs, 5 to 8 spines, the number of spines will increase while growing; spines yellow or brown; flowers bright purple-red, might be the most beautiful flowers in family Cactaceae; native to central Mexico.

黄金纽属与仙人球属 – 杂交种
Hildewintera X Echinopsis

植株细长，具有黄金纽特征，花红色，花柄较长，具有仙人球特征。

Whole plant thin and long; flowers red with long flower stalk.

量天尺
Hylocereus undatus (Haw.) Britton & Rose

植株多为三角形，多节，新生刺座具 1~3 根黑色或褐色刺。花内瓣白色，外瓣黄绿色，可供食用。其花俗称霸王花。

Usually triangular; newborn areole has 1 to 3 black or brown spines; flower inside petals white, outside petals yellowish green; eatable.

光山
Leuchtenbergia principis Hook.

植株形态似龙舌兰，故称龙舌兰仙人掌，高达 30 cm，疣突三棱形，蓝绿色，细长，约 10~12 cm，周刺 8~14 根，白色或黄褐色，长约 10~15 cm，花黄色，有香气。原产墨西哥中北部。

Whole plants look like maguey, 30 cm tall; flowers yellow, aromatic; native to north central Mexico.

黄裳
Lobivia aurea (Britton & Rose) Backeb.

植株球形，有 14~16 条棱，植株灰绿色，周刺 8~12 根，白色，中刺 1~3 根，黑色，花柄长 4~6 cm，花大，黄色。原产阿根廷。

Whole plant spherical, greyish green; 14 to 16 ribs; 8 to 12 white outer spines, 1 to 3 black central spines; flower stalk 4 to 6 cm long; flowers big and yellow; native to Argentina.

黄裳（变种）
Lobivia aurea var. fallax (Oehme) Rausch

植株球形，绿色，长有8~12条棱，周刺8~10根，白色或淡灰褐色，长约2~3 cm，中刺1~3根，黑色。植株顶部中刺棕黑色，长约4~6 cm，花金黄色。原产阿根廷。

Whole plant spherical, green; 8 to 12 ribs; 8 to 10 white or pale grey outer spines, 2 to 3 cm long, 1 to 3 black central spines, 4 to 6 cm long; flowers golden yellow; native to Argentina.

湘阳球
Lobivia bruchii Britton & Rose

植株球形，在基部萌生出子球，丛生。30~32条带疣突纵向短棱。刺座长于疣突节上，具6~8枚白色针状周刺和1枚较长的针状中刺。花红色。原产阿根廷。

Whole plant spherical, cormels grow at base, clustered; 30 to 32 short ribs; 6 to 8 white, needle like outer spines, 1 long, needle like central spine; flowers red; native to Argentina.

凯南娜丽花球
Lobivia caineana Cárdenas (Echinopsis) c. (Cardenas) D.R. Hunt

植株圆筒形，有10~12条较宽阔的棱，刺座长于棱缘上。花顶生，粉红色。原产玻利维亚。

Whole plant terete; 10 to 12 broad rids, areoles grow on the edges of these rids; flowers pink and acrogenous; native to Bolivia.

密刺丽花球（黄花）
Lobivia densispina Werderm. ex Backeb. & F. M. Knuth

植株绿色或紫褐色，圆柱形，生有直棱，中刺4~7根，刺短，黄色或褐色齿形排列，花侧生，浅黄色，红色或紫色；花喉部色深。原产阿根廷。

Whole plants green or purple-brown, terete; ribs straight; 4 to 7 short, yellow or brown central spines; flowers pale yellow, red or purple; native to Argentina.

华宝球

Lobivia einsteinii (Frić) Rausch

华宝球

Lobivia einsteinii (Frić) Rausch

植株多分枝，纯绿色，上有红色或深褐色斑点，长条形或球形，粗2 cm，具13~16条直或斜棱，周刺12根，花深黄色，喉部则变淡至黄色。原产阿根廷。

Whole plant green with red of dark brown spots, strip shaped and spherical, diameter 2 cm, multi-branched; 13 to 16 ribs; 12 outer spines; 12 spines; flowers dark yellow; native to Argentina.

阳盛球
Lobivia famatimensis (Speg.) Britton & Rose

植株初生球上有密生刺，紧贴球体开喇叭型黄色花。原产阿根廷。

Spines dense; flowers trumpet like, yellow; native to Argentina.

朱丽球
Lobivia hertrichiana Backeb.

植株单生或分支，球形或椭圆形，生有11~12条带疣突的直棱或斜棱，周刺5~14根，长约2 cm，中刺1~7根，长约2.5 cm，花侧生，橘红色、朱红色或粉红色。原产秘鲁。

Whole plant solitary or branched, spherical or oval; 11 to 12 ribs; flowers orange-red, vermilion or pink; native to Peru.

红笠球
Lobivia jajoiana Backeb.

红笠球（黄花）
Lobivia jajoiana Backeb.

植株多单生，椭圆形或圆柱形，生有 10~18 条呈疣突的斜棱。周刺 8~11 根，长 1~3 cm，中刺 1~3 根，长 1.5~6 cm，其中一根向上生长，有钩，花侧生，具香气，黄色，橘红色，酒红色或深西红柿色，花喉部色深，黑色。原产阿根廷。

Whole plants usually solitary, oval or terete; 10 to 18 ribs; 8 to 11 outer spines, 1 to 3 cm long, 1 to 3 central spines, 1.5 to 6 cm long; flowers yellow, orange-red, wine-red or dark red, aromatic; native to Argentina.

布氏巨黄丸丛生变种
Lobivia maximiliana subsp. caespitosa (J.A. Purpus) Rausch ex G.D. Rowley

花为橙色。分布于玻利维亚。

Flowers orange. Distributed in Bolivia.

青玉
Lobivia pentlandii (Hook.) Britton & Rose

植株单生或分支，球形或椭圆形，生有10~20条有疣突的斜棱。周刺6~15根，长0.4~4.5 cm，中刺0~1根，长2~9 cm，花侧生，黄色，橘红色，朱红色，浅红或粉红色；也有黄白色或紫粉色的，花非常多样。原产秘鲁和玻利维亚北部。

Whole plant solitary or branched, spherical or oval; 10 to 20 ribs; 6 to 15 outer spines, 0.4 to 4.5 cm long, 0 to 1 central spines, 2 to 9 cm long; flowers yellow, orange-red, vermilion, pale red or pink, sometimes yellowish white or purple-pink; native to Peru and northern Bolivia.

姬丽球（变种）
Lobivia schieliana var. leptacantha (Rausch) Rausch

植株分枝，粗 1.5~3 cm，具 9~14 条棱，周刺 6~8 根，无中刺，偶有 1 根，花紫红色。原产阿根廷。

Whole plant branched, diameter 1.5 to 3 cm; 9 to 14 ribs; 6 to 8 outer spines, no central spine; flowers purple-red; native to Argentina.

白檀
Lobivia silvestrii (Speg.) G. D. Rowley

植株分枝多，手指粗细，具 6~10 条直棱，周刺 10~15 根，刚毛状，0.2 cm 长。无中刺。花侧生，朱砂红色。原产阿根廷。

Whole plant multi-branched; 6 to 10 straight ribs; 10 to 15 bristle like outer spines, 0.2 cm long, no central spines; flowers vermilion; native to Argentina.

白檀杂交种
Lobivia silvestrii (Speg.) G.D. Rowley (Hybrid)

牡丹球
Lobivia winterina F. Ritter

植株单生，灰绿色，多年生后变长形，具 13~19 条有突起的斜棱，周刺 6~14 根，向植株弯曲生长，中刺 0~1 根，部分有钩。花侧开，宝石红色，花茎 7~9 cm，相比植株花很大。原产秘鲁。

Whole plant solitary, greyish green; 13 to 19 verrucose ribs; 6 to 14 outer spines, 0 to 1 hooked central spines; flowers ruby, relatively big; native to Peru.

桃轮球
Lobivia wrightiana Backeb.

植株多为单生，椭圆形，有 12~17 条有突起的斜棱，周刺 6~10 根，长 0.5~3 cm，中刺多为 1 根 1~4 cm 长，刺尖有钩，这是一种颇受栽种者欢迎的种类，花粉红色，早春开花。原产秘鲁中部。

Whole plant usually solitary, oval; 12 to 17 ribs; 6 to 10 outer spines, 0.5 to 3 cm long, 1 hooked central spine, 1 to 4 cm long; flowers pink, florescence is in early spring; native to central Peru; native to middle Peru; deeply favored by plants enthusiasts.

黑乳突球
Mammillaria backebergiana F. G. Buchenau

花为紫红色呈环状坐于植株顶部。分布于墨西哥。

Flowers purplish red, in double or triple rings at the plant apex. Distributed in Mexico.

卡氏乳突球
Mammillaria casoi Bravo

球形，花小为红色，呈环状在植株顶部。分布于墨西哥瓦哈卡州。
Spherical. Flowers small, red, in ring at the top of plant apex. Distributed in Oaxaca, Mexico.

雪月花
Mammillaria elegans DC.

植株圆柱形，具环节，周刺16~29根，雪白色，紧贴球体，中刺2根，短。花小，红色，沿植株顶呈环状着生。原产墨西哥。
Whole plant terete; 16 to 29 white outer spines, 2 short central spines; flowers small and red; native to Mexico.

林氏乳突球
Mammillaria hubertller 'Reppenhagen'

金星
Mammillaria longimamma DC.

植株单生或分支,球形,无乳汁,疣突,有毡毛或光秃,周刺9~10根,白色或淡黄色,无中刺或1根,花为浅黄色,花朵紧贴茎部开放,种子黑色。原产墨西哥。

Whole plant solitary or branched, spherical; 9 to 10 white or pale yellow outer spines, 0 to 1 central spine; flowers pale yellow, grow on stems; native to Mexico.

柏氏乳突球

Mammillaria parkinsonii Ehrenb.

丛生。分布于墨西哥的伊达尔戈州。
Clumping. Distributed in Hidalgo, Mexico.

月宫殿

Mammillaria senilis Lodd. ex Salm-Dyck

植株球形或卵形，无乳汁，疣突有白色绒毛或刚毛，周刺30~40根，黄白色或雪白色，中刺5~6根，刺尖黄褐色。花橘红色或紫色，带有深色中分条纹，有鳞片。原产墨西哥。

Whole plant spherical or oval; warts covered with white tomentum or bristle; 30 to 40 yellowish white or white outer spines, 5 to 6 central spines with brown apex; flowers orange-red or purple with dark stripe; native to Mexico.

无刺乳突球
Mammillaria theresae Cutak

植株多单生，卵形，无乳汁，疣突有绒毛，周刺20~30根，中刺9根，刺均呈羽毛状，花为紫色或紫红色，长3.5~4.5 cm，种子黑色。原产墨西哥。

Whole plant solitary, oval; warts covered with tomentum; 20 to 30 outer pines, 9 central spines, all of them are feather like; flowers purple or purple-red; native to Mexico.

黄仙玉
Matucana aurantiasa (Vaupel) Buxb.

植株球形至长球形，16条扁平疣突状的棱，刺10~25根，黄色或红褐色，花橘红色，向心变橘黄色。原产秘鲁北部。

Whole plant spherical or long spherical; 16 flat, verrucose ribs; 10 to 25 spines, yellow or red brown; flowers orange-red; native to northern Peru.

帕拿利白玉仙
Matucana pallarensis F. Ritter

植株黄绿色至绿色，具 14~18 条不深的直棱，刺座偏长，周刺 16~20 根，白色，或淡金黄色，花橙黄色，棒状，高 6~8 cm。原产秘鲁。

Whole plants yellowish green to green; 14 to 18 ribs; 16 to 20 white or pale golden outer spines; areole relatively long; flowers orange, rodlike, 6 to 8 cm tall; native to Peru.

白头花座球
Melocactus albicephalus Buining & Brederoo

植株单生，绿色，球径 12~15 cm，有 9~10 条突起纵向棱。花座长满白色短刚毛。花朱红色，小钟状。原产中美洲和加勒比海岛屿。

Whole plant solitary, green, diameter 12 to 15 cm, 9 to 10 ribs; cephalium covered with white, short bristles; flowers vermilion, small campanulate; native to Central America and Caribbean Islands.

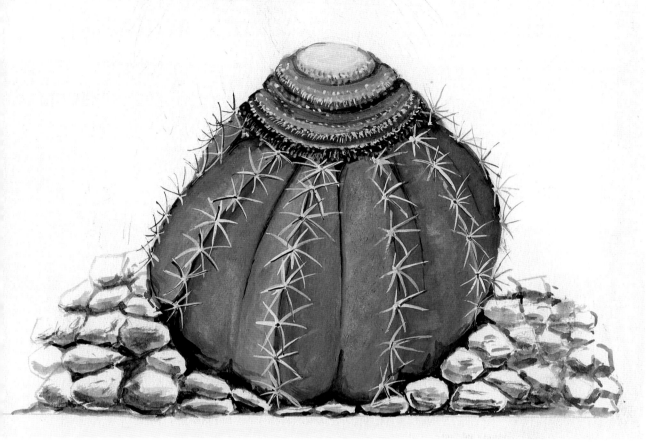

蓝云
Melocactus azureus Buinging & Brederoo

植株单生，具蓝色霜斑以至植株呈灰蓝色，球径超过 14 cm。具 9~10 条不算尖的棱。花座具白色绒毛和红色刚毛，呈环状交错生长。原产巴西。

Whole plants solitary, greyish blue, diameter more than 14 cm; 9 to 10 ribs; native to Brazil.

赫云
Melocactus bahiensis (Britton & Rose) Luetzelb.

植株单生，在原生地偶有群生，灰绿色至黄绿色，扁球形至球形。具 13~15 条高背直棱。具 12~14 枚灰白色放射状周刺和 2~4 枚针状中刺。花座上白色毛和深枣红色刚毛成环状交错生长。花座顶以白色为主，花小，洋红色，着生于花座顶。原产西印度群岛。

Whole plant solitary, rarely clustered, greyish green to yellowish green, flat spherical to spherical; 13 to 15 straight ribs; 12 to 14 greyish white, ridial outer spines, 2 to 4 needle like central spines; cephalium covered with white and purplish red bristles; flowers small, carmine, grow on top; native to the West Indies.

层云
Melocactus bellavistensis Rauh & Backeb.

植株单生。在原生地有群生。具 10~12 条尖厚纵向棱。具 6~8 枚灰白色针状周刺和 1 枚较长褐色针状中刺。花座为紫红色，与白色短刚毛交替生长。花桃红色，小钟状，环开于花座上。果实为小香肠形，洋红色，留在花座上的时间较长。原产哥伦比亚、巴西等地。

Usually solitary, rarely clustered; 10 to 12 sharp, thick, sharp ribs; 6 to 8 greyish white, needle like outer spines, 1 long, brown, needle like central spine; cephalium purple-red; flowers peach red, little campanulate; native to Columbia and Brazil.

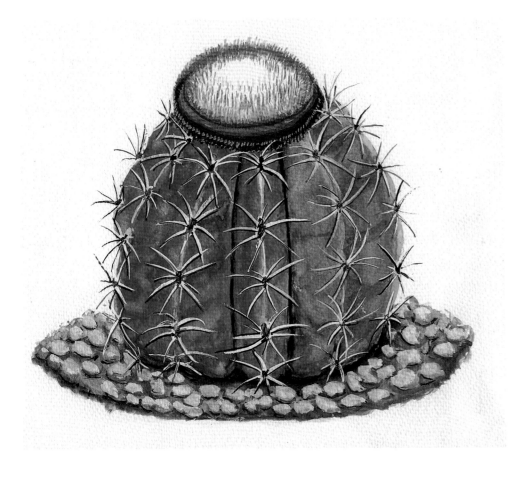

爪刺花座球
Melocactus bellavistensis subsp. onychacanthus (F. Ritter) N.P. Taylor

植株卵形，深绿色，长有 8~10 条扁平宽棱，6~10 根白色贴球体生长周刺，长约 3~5 cm，植株顶部较扁平，边缘长有棕红色密密的短刚毛，花小，红色，果实红色小香肠形。原产巴西等地。

Whole plants oval, dark green; 8 to 10 flat broad ribs; 6 to 10 white outer spines, 3 to 5 cm long; flowers small and red; fruits red, sausage shaped; native to Brazil.

彩云
Melocactus bellavistensis Rauh & Backeb.

花座高于植株顶部呈红白相隔环状。分布于墨西哥。

Cephalium covered with white hairs interspersed with red bristles. Distributed in Mexico.

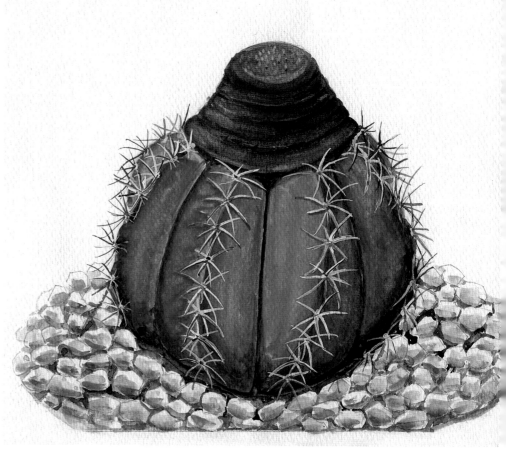

姬云
Melocactus concinus Rauh & Backeb.

植株单生，具 10~13 条尖锐纵向直棱。刺座着生于棱脊上，具 6~8 枚灰白色向内微弯扁刺。花座为白色和红棕色短刚毛组成。花朱红色，较易开花。原产巴西。

Whole plant solitary, 10 to 13 sharp straight ribs; 6 to 8 greyish white, flat spines slightly curve inward; cephalium formed by white and reddish brown bristles; flowers vermilion; native to Brazil.

格氏花座球
Melocactus caesius H.L. Wendl.

植株卵形，顶上生有塔形花座，灰绿色，长有 10 条扁平棱，周刺黑色，8~10 根，贴球体生长，花小红色。原产于委内瑞拉和西印度群岛。

Whole plants oval, 10 flat arrises; flowers small and red; native to Venezuela and the West Indies.

巴哈马花座球
Melocactus ernestii Vaupel

植株长球形，顶部花座球较高，有棕黑色刚毛圈。顶部多为白色刚毛，花红色细小，开在花座球顶部，茎橄榄绿色，具 12~14 条粗阔厚直的棱，刺座深褐色至黑色，周刺 10~16 根淡褐色，中刺 3~6 根淡褐色，粗壮，向下生长，长约 4~5 cm。原产西印度巴哈马。

Whole plant long spherical; white bristle grow on top; flowers red and small, grow on top; stems olive green; 12 to 14 broad, thick and straight ribs; 10 to 16 pale brown outer spines, 3 to 6 thick, pale brown central spines, 4 to 5 cm long; areole dark brown or black; native to west Indian, Bahamas.

马坦萨斯花座球
Melocactus matanzanus León

乱云
Melocactus oreas Miq.

植株单生，深绿或浅绿色，球径粗 12~22 cm，具 10~16 条高背纵向直棱。花座白色和红棕色绒毛与刚毛交替组成。原产巴西东部。

Whole plant solitary, dark green or pale green, diameter 12 to 22 cm; 10 to 16 staight ribs; cephalium formed by white and reddish brown tomentum and bristles; native to eastern Brazil.

华云
Melocactus peruvianus Vaupel

花座生于植株顶部，细刺为红色，细刺成环状包围，花小为红色，果实为红色小条状。广布于秘鲁。

Cephalium around with red spines. Flowers small, red. Fruits red like small strip. Widespread in Peru.

碧云
Melocactus salvadorensis Werderm.

植株单生，扁球形至球形，球径约 10 cm。花座着生密集白色短刚毛，偶尔有小量红棕色短刚毛，交错组成。生长较缓慢，约 10 年才能开花结果。原产巴西。

Whole plants solitary, flat spherical to spherical, diameter 10 cm; cephalium covered with short, white bristles; native to Brazil.

残雪
Monvillea spegazzinii (F.A.C. Weber) Britton & Rose

植株灌木状，为攀附型柱状仙人掌，茎有白斑似残雪。具5~12条棱，花生在茎中部或基部，花管较长，夜间开花。原产于南美洲巴西、巴拉圭等地。

Shrubs, epiphytic and terete; stems covered by white spots; 5 to 12 ribs; flowers grow in the middle or base of stems; flowers nocturnal, floral tubes long; native to Brazil and Paraguay.

红花大凤龙（勇凤）
Neobuxbaumia euphorbioides Buxb.

植株为浅灰蓝绿色；高达数米，株幅6~9 cm，具细长尖棱8~10条，刺1~5根。花侧开，初开黑色后渐为红至粉红。生长较粗犷，值得栽种。原产墨西哥。

Whole plant pale grey and blue-green; up to several meters high, 6 to 9 cm wide; 8 to 10 ribs, 1 to 5 spines; early blossoms black, gradually turn into red and finally pink; suitable for cultivation; native to Mexico.

圆锥玉
Neolloydia conoidea (DC.) Britton & Rose

植株高 8~10 cm, 株幅约 4 cm, 不易分枝。刺座有白色绒毛，周刺 16~25 根，白色，紧贴在茎棱上。中刺 1~3 根，黑色。花顶生，深紫红色。原产美国南部和墨西哥。

Whole plant 8 to 10 cm high, 4 cm wide; not easy to branch; white tomentum grow on areole; 16 to 25 white spines tightly grow on stem ribs; 1 to 3 black spines; flowers grow on top, dark purple-red; native to southern U.S. and Mexico.

智利多色球（豹头）
Neoporteria napina (Phil.) Backeb.

花为白色到粉色、黄色或棕色。分布于智利的科皮亚波省。

Flowers white to pink, yellow or brownish. Distributed in Copiapo, Chile.

玉姬
Neoporteria sp.

英冠玉（壮丽南国玉）
Notocactus magnificus (F. Ritter) Krainz ex N.P.Taylor

植株早期单生，生长到一定阶段开始分枝，群生植株球形至长球形，球径 12~14 cm，高约 20 cm，亮绿色。具 12~16 条光滑、较深的锐棱。刺座长于棱缘上，并长有白色绒毛，具 12~15 枚金黄色丝发状的周刺。花群生，亮黄色，顶生。花期夏季，较长。原产巴西。

Newborn plants solitary, gradually branched and clustered, spherical to long spherical, diameter 12 to 14 cm, 20 cm tall, bright green; 12 to 16 smooth, sharp ribs; areole grow on rib margin, covered with white tomentum; 12 to 15 golden, hair like outer spines; flowers clustered, bright yellow, grow on top; bloom in summer; native to Brazil.

壮丽南国玉
Notocactus magnificus (F. Ritter) Krainz ex N.P.Taylor

花为浅黄色。分布于巴西南里奥格兰德州。

Flowers pale yellow. Distributed in Rio Grande do Sul, Brazil.

短尖刺南国玉
Notocactus muricatus (Otto) Backeb.

植株一般丛生，球形至椭圆形，绿色至黄绿色。具18~22条具斧突纵向直棱。具5~7枚短硬针状周刺和1枚中刺。花顶生，亮黄色。花蕊枣红色。花柄为褐色绒毛所包裹。原产巴西。

Whole plant usually solitary, spherical to oval, green to yellowish green; 18 to 22 straight ribs; 5 to 7 short, hard, needle like outer spines, 1 central spine; flowers acrogenous and bright yellow; flower stalk covered with brown tomentum; native to Brazil.

拉氏南国玉
Notocactus rauschii Vliet

球形，花大为黄色。分布于乌拉圭。
Globular. Flowers large, yellow. Distributed in Uruguay.

眩美玉
Notocactus uebelmannianus Buining

花为紫色或黄色。分布于巴西南里奥格兰德州。
Flowers purple or yellow. Distributed in Rio Grand do Sul, Brazil.

范氏南国玉
Notocactus vanvlietii Rausch

花为黄色到橙黄色。分布于乌拉圭。
Flowers yellow to orange yellow. Distributed in Uruguay.

大花南国玉（黄花）
Notocatus Hybrid

周刺 8~12 条，较粗壮，花硕大，黄色，为一栽培种，颇受种植者喜爱。

A cultivar; 8 to 12 thick outer spines; flowers big and yellow; deeply favored by plants enthusiasts.

南国玉 – 杂交栽培种
Notocatus Hybrid

未被鉴定南国玉
Notocatus sp. (Hybrid)

植株扁球形，具 22~26 条浅棱，刺座较密，分枝，花鲜红色，硕大。
Whole plant flat spherical; 22 to 26 ribs; areole dense; branched; flowers big and bright red.

南国玉
Notocatus sp. (Hybrid)

帝冠
Obregonia denegrii Fric

这个属只有一种,植株扁平,形似皇冠因此得名,花白色或浅粉红,生长于富含矿物质的土壤,半阴环境。

There is only one species in this genus; whole plant flat, looks like a crown; flowers white or pale pink; grow in mineral-rich soil.

毕氏团扇(红花团扇)
Opuntia elatior Mill.

植株掌状,属节刺类仙人掌,刺座灰褐色,周刺10~14根,花黄色,橙色和红色。原产墨西哥。

Whole plant palmlike; areole greyish brown, 10 to 14 spines; flowers yellow, orange or red; native to Mexico.

伯氏仙人掌
Opuntia bergeriana F.A.C. Weber ex A.Berger

分布于美国北部、南部和中部及西印度群岛。

Distributed near all parts of north, central and south America, also West Indies.

松岚
Opuntia bigelovii Engelm.

植株多分枝，呈小灌木状。每节长满白色针状刚刺，短而整齐，具光泽，十分美观。花为绿白色。原产加利福尼亚州南部，亚利桑那州西部和南部，墨西哥下加利福尼亚州等。

Whole plant muti-branched, small shrub like; flowers green-white; native to southern California, western and southern Arizona, U.S. and Mexico.

红点团扇
Opuntia gosseliniana F.A.C. Weber

植株多分枝，群生，茎节扁圆形，形似乒乓球拍，灰黄绿色至灰绿色。刺座长有红色或棕红色一撮钩芒刺。花黄色。原产美国南部和墨西哥加利福尼亚半岛。

Whole plant multi-branched, clustered; flowers yellow; native to southern U.S. and California peninsula, Meixco.

长网孔仙人掌
Opuntia longiareolata Clover & Jotter

花为深玫瑰色。分布于美国加利福尼亚州南部、内华达州南部、亚利桑那州西部和犹他州南部。

Flowers deep rose. Distributed in southern California, southern Nevada, western Arizona, and southern Utah, the United States.

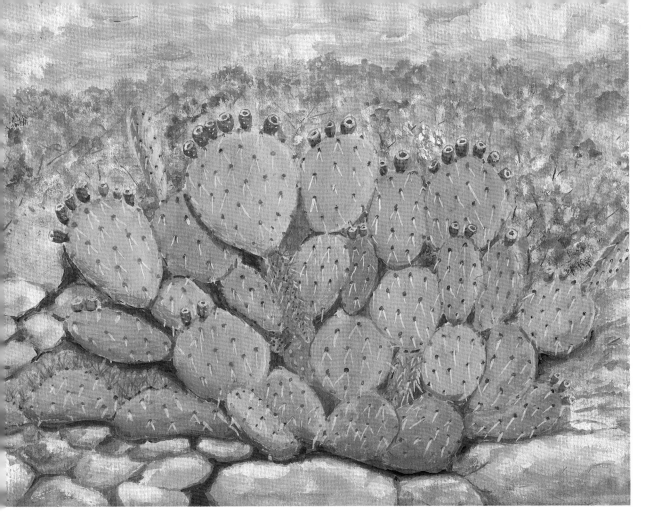

褐刺仙人掌
Opuntia phaeacantha Engelm.

花为柠檬黄色,花期在春末。分布于美国亚利桑那州中部和南部。

Flowers lemon-yellow, florescence is in the late spring. Distributed in central and southern Arizona, the United States.

仙人镜
Opuntia robusta J.C. Wendl.

植株多分枝,具主茎,树状。茎节与一般团扇不同,呈长椭圆形,像美人镜,绿色至灰绿色。刺细小且短,数量不多。花橙色,秋季开放。果实红色。原产墨西哥中部山区。

Whole plant multi-branched, tree like; stem nodes long oval, like mirrors, green to greyish green; spines tiny and short; flowers orange, bloom in autum; fruits red; native to central Mexico.

红团扇（淡褐刺仙人掌）
Opuntia rufida Engelm.

花为鲜黄色。分布于美国得克萨斯州西部和墨西哥奇瓦瓦州及科阿韦拉州。
Flowers light yellow. Distributed in western Texas, the United States and Chihuahua, Coahuila, Mexico.

萨氏团扇
Opuntia salmiana Parm. ex Pfeiff

植株主茎光滑，无明显分枝，刺座灰白，枝茎有明显的棱，生有2~3根刺，红色，叶茎生有12根长刺。原产玻利维亚。
Stems smooth, non-branched; areole greyish white; several ribs, 2 to 3 spines; leaf stems 12 long spines; native to Bolivia.

鹤岑球
Oreocereus doelzianus (Backeb.) Borg

植株茎长1米，粗4~8 cm，具9~11条棱，周刺10~16根，中刺4根，刺黄色或深褐色，花红色，10 cm长。易于生长。原产秘鲁中部。

Stems 1 m high, diameter 4 to 8 cm; 9 to 11 ribs; 10 to 16 outer spines, 4 central spines, all of them yellow or dark brown; spines yellow or dark brown; flowers red, 10 cm long; easy to cultivate; native to middle Peru.

圣云龙
Oreocereus hendriksenianus Backeb.

分布于秘鲁高海拔地区。为毛柱类仙人掌一种。
Clumping. Distributed in high altitude area in Peru.

武烈丸
Oreocereus neocelsianus Backeb.

植株球形或长球形，高 20 cm，株幅 40 cm。35 条棱，刺白色、黄色、红色或黑色。周刺约 10~30 根，齿状。中刺 0~6 根不等。花浅朱红色或朱砂红色，花喉部黄色或柠檬黄色。原产秘鲁中部。

Whole plant spherical or long spherical; 20 cm high, 40 cm wide; 35 ribs, spines white, yellow, red or black; flowers pale or dark vermillion; native to middle Peru.

红花天轮柱
Pachycereus schottii (Engelm.) D.R. Hunt

植株为巨大乔木状柱形仙人掌，具 10~12 条棱，茎分叉，灰绿色，花不大，红色。原产墨西哥。

Whole plant tree like and terete; 10 to 12 ribs; stems branched and greyish green; flowers small and red; native to Mexico.

玻利维亚锦绣球
Parodia sp.

拜氏南国玉
Parodia buiningii (Buxb.) N. P. Taylor

花为浅黄色，花期在 10~11 月。分布于乌拉圭。
Flowers pale yellow, appearing in October-November. Distributed in Uruguay.

海王球
Parodia crassigibba (F. Ritter) N.P. Taylor

株幅5~16cm，10~15条棱，顶端秃，无刺，周刺7~10根，中刺最多1根，刺为浅褐色发灰，长3cm，花黄色顶生，原产巴西。

Whole plants 5 to 16 cm wide, 10 to 15 ribs; 7 to 10 outer spines, 0 to 1 central spines, all of the spines are pale brown; flowers acrogenous; native to Brazil.

硬刺锦绣球
Parodia echinopsoides F. H. Brandt

花为橙红色，刺金黄色，生于植株顶端。分布于玻利维亚。

Spines gold yellow, flowers orange-red at the top of the plant. Distributed in Bolivia.

赫氏南国玉
Parodia horstii (F. Ritter) N.P. Taylor

植株球形，具6~15条棱，刺座无毛，周刺10~16根。花红色，顶生，原产自南美洲乌拉圭、巴拉圭和阿根廷。

Whole plant long spherical; 6 to 15 ribs, areole hairless, 10 to 16 spines; flowers red and acrogenous; native to Uruguay, Paraguay and Argentina.

黄翁
Parodia leninghausii (Haage) F. H. Brandt

花为黄色，花期是7~8月。分布于巴西南里奥格兰德州。

Flowers yellow, appearing in July-August. Distributed in Rio Grande do Sul, Brazil.

红刺魔神
Parodia maassii (Hesse) A. Berger

植株球形或长球形，株高 20 cm，株幅 15 cm。具 20~22 条稍斜的直棱，周刺 6~18 根，棕红色。中刺 1~6 根，末端有钩，长 7 cm。花黄色，顶生，不大。原产玻利维亚南部。

Whole plant spherical or long spherical, 20 cm tall, 15 cm wide; 20 to 22 ribs; 6 to 18 reddish brown outer spines, 1 to 6 hooked central spines, up to 7 cm long; flowers small and yellow and acrogenous; native to south Bolivia.

黑云龙
Parodia maassii var. subterranea (F. Ritter) Krainz

植株球形，株幅约 6 cm，茎顶白色，周刺约 10 根，白色角质，刺突呈黑色，中刺 1~4 根，黑色，花宝石红色或紫红色。原产玻利维亚。

Whole plant spherical; 6 cm wide; stem top white; 10 outer spines, 1 to 4 black central spines; flowers ruby red or purple-red; native to Bolivia.

金冠
Parodia schumanniana (Nicolai) F. H. Brandt

植株初生为单生，后开始分枝成群生，球形至长球形，绿色至黄绿色。具12~14条较深纵向直棱。刺座生于棱缘上，并长有20~24枚金黄色毛发状周刺。花亮黄色，顶生。花期夏季，开花时间较长。原产巴西。

Newborn plants solitary, gradually branched and clusterd, spherical to long spherical, green to yellowish green; 12 to 14 straight ribs; areoles grow on ribs margin; flowers bright yellow and acrogenous; bloom in summer; native to Brazil.

红色具刺南国玉
Parodia neoarechavaletae (Havlicek) D.R. Hunt.

花为红色生于植株顶部并有长刺包围。分布于乌拉圭的马尔多纳多省附近。

Flowers red, growing at the top of the plant, around with long spines. Distributed near Maldonado, Uruguay.

蛇状丝柱
Peniocereus viperinus (F.A.C. Weber) Buxb.

植株长条状，有分枝，茎褐色，具白色绒毛的刺座。花红色，罗伞形，较密。原产墨西哥。

Whole plant long strip-shaped, branched; stems brown; areole white and villiform; flowers red and dense; native to Mexico.

白眉塔
Peniocereus viperinus var. (F.A.C. Weber) Buxb.

植株长条形，花开在茎中，盛开的鲜红色，像拟萝伞。具白绒毛状刺座。原产于墨西哥。

Whole plant strip-shaped; flowers bright red and grow in stems; areole white and villiform; native to Mexico.

市麒麟
Pereskia bahiensis Gürke

植株具有茎，叶，枝和针状刺；红枣色，攀援生长。原产南美洲巴西、巴拉圭。
Whole plant have stems, leaves, branches and needlelike spines; red and climbing; native to Brazil and Paraguay.

叶花仙人掌
Epiphyllum (Hybrid)

叶花仙人掌多为杂交栽培种，所以花大且多，颜色多样，为栽培者喜爱。
A cultivar; flowers big and colorful; deeply favored by plant enthusiasts.

薇拉娜
Phyllocactus 'Verana' (Hybrid)

黄叶掌
Epiphyllum (Hybrid)

普氏叶掌杂交种
Epiphyllum (Hybrid)

乌夫人
Epiphyllum 'Frau H. Wegener' (Hybrid)

五月草
Epiphyllum 'Mae Marsh' (Hybrid)

红雀
Epiphyllum 'Red Bird' (Hybrid)

红叶掌杂交种
Epiphyllum 'Red' (Hybrid)

洋娃娃马狄逊
Epiphyllum 'America Sweetheart' (Hybrid)

美国情人
Epiphyllum 'America Sweetheart' (Hybrid)

狮子
Pilosocereus leucocephalus (Poselger) Byles & G. D. Rowley

植株树形，高达6米，茎为深绿色，具7~9条浑圆的棱。刺座着生有白色棉毛状绒毛。周刺7~12根，中刺1根长3cm。全部的刺均为褐色或灰色，花紫红色，夜间开放，越冬不能低于15℃。原产墨西哥。

Whole plants tree like, up to 6 m high; stems dark green; 7 to 9 round ribs; white tomentum grow on the areole; spines brown or grey; flowers purple-red and nocturnal; native to Mexico.

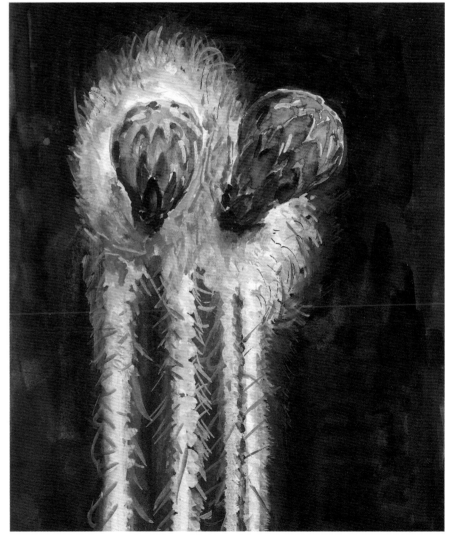

伯氏毛柱
Pilosocereus pachycladus F. Ritter

植株高大，可达 10 米，蓝绿色有 4 条深厚棱，棱缘生有白色绒毛，周刺 4~6 根，中刺一根较长，白色。原产巴西。

Whole plants huge and tall, up to 10 m; 4 blue-green ribs; 4 to 6 outer spines, 1 white, long central spine; white tomentum grow on arris margin; native to Brazil.

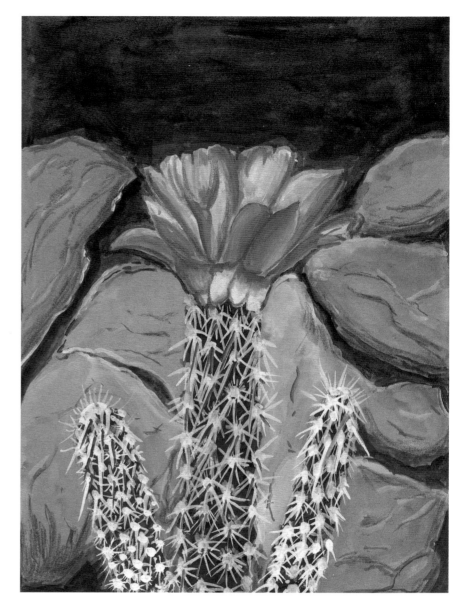

直刺仙人掌
Pterocactus fischeri Britton & Rose

植株为柱形仙人掌，生有 3~4 条翅状棱，花靠顶侧生，外褐色，内淡褐色，外有鳞片。刺毡毛状，斜向下生直刺，长 6 cm，白色。原产阿根廷。

Whole plant terete; 3 to 4 wing like ribs; flowers outside brown, inside pale brown; spines white, straight, up to 6 cm long and grow downward; native to Argentina.

网脉有翼柱
Pterocactus reticulatus R. Kiesling

植株柱形，生有螺旋状的斜棱，周刺 10~16 根，白色，花顶生，淡粉红色。原产阿根廷。

Whole plant terete with spiral ribs; 10 to 16 white spines; flowers pale pink and acrogenous; native to Argentina.

黑龙
Pterocactus tuberosus (Pfeiff) Britton & Rose

植株红褐色，为一小型种类，12 cm 高，茎粗 1 cm，花黄色，顶生。原产阿根廷。

Whole plant red-brown, a small species, 12 cm tall, stems diameter 1 cm; flowers yellow and acrogenous; native to Argentina.

真春黄菊子孙球
Rebutia aureiflora Backeb.

植株多分枝，茎灰色或橄榄绿，长形，约生有10~14条棱，周刺约12根，白色，花朱红色。原产阿根廷。

Whole plant multi-branched; stems grey or olive green; strip shaped; 10 to 14 ribs; 12 white spines; flowers vermilion; native to Argentina.

珠节子孙球
Rebutia torquata F. Ritter & Buining

植株长形，茎深绿色，8~10条不规则的棱，刺座较密，布满棕色短刺，花橙色，花蕊渐变黄色，花生于茎部。原产阿根廷。

Whole plant strip shaped; stems dark green; 8 to 10 irregular ribs; brown short spines grow densely on the areole; flowers orange, grow on the stem; native to Argentina.

紫宝球
Rebutia violaciflora Backeb.

植株多单生,黄绿色,刺金褐色,花浅紫色,周刺长 0.3~2.5 cm。原产阿根廷。
Usually solitary; whole plants yellowish green; spines golden brown; flowers pale purple; spines 0.3 to 2.5 cm long; native to Argentina.

优雅（仙人指杂交种）
Schlumbergera 'Delicatum' (Hybrid)

黄金蟹（仙人指杂交种）
Schlumbergera 'Goldcharm' (Hybrid)

仙人指杂交种
Schlumbergera 'Le Vesuv' (Hybrid)

仙人指属杂交种
Schlumbergera Hybrid

园辨仙人指（锦上添花）
Schlumbergera truncata (Haw.) Moran

植株为叶状茎，边缘有 2~4 个锯齿状刺，宽 1.5~2.5 cm，长 4.5~5.5 cm。刺座有微小毡毛，具 1~3 根短刚毛，花对称，粉红色或深紫色。原产巴西。

Leaflike stems, 2 to 4 zigzag spines grow on the margin; 1.5 to 2.5 cm wide, 4.5 to 5.5 cm long; tiny tomentum grow on areole; flowers pink or dark purple, symmetrical; native to Brazil.

大花蛇鞭柱（大轮柱）
Selenicereus grandiflorus (L.) Britton & Rose

植株长条状，茎具 5~8 条棱，宽 2~3 cm，刺座有白色或黄色棉毛，初生有 7~11 根针状刺，外轮花瓣橙红色，内轮花瓣白色，花大，达 18~30 cm，具香味。原产墨西哥，海地，牙买加，古巴等。

Whole plant strip shaped; 5 to 8 ribs, 2 to 3 cm wide; white or yellow tomentum grow on areole; newborns have 7 to 11 acicular spines; outside petals orange-red, inside petals white, flowers diameter 18 to 30 cm; native to Mexico, Haiti, Jamaica and Cuba.

夜美人
Selenicereus pteranthus (Link ex A. Dietr.) Britton & Rose

植株长条状，攀援生长，茎具 4~5 条棱，青绿色，粗 2~5 cm，刺座生有白色棉状绒毛，有 6~12 根由黄色变为灰色，短小圆锥形刺，长度仅为 0.6 cm。花外瓣红黄色，内瓣白色或奶黄色。原产墨西哥。

Whole plant strip shaped and climbing; 4 to 5 ribs, pale green, diameter 2 to 5 cm; white tomentum grow on areole; 6 to 12 short conical spines, turning from yellow to grey; outside petals red-yellow, inside petals white or cream yellow; native to Mexico.

奇想球
Setiechinopsis mirabilis Backeb. ex de Haas

植株大多不分枝，为一小型种类，株高 10~15cm，株幅 2~2.5cm，具 11~12 条棱；花顶生；种子繁殖，极易开花。原产阿根廷。

As small species, rarely branched; 10 to 15 cm tall, 2 to 2.5 cm wide; 11 to 12 ribs; flowers acrogenous; native to Argentina.

菊水
Srombocactus disciformis (DC.) Britton & Rose

植株扁平具棱形疣突球，其刺易脱落，花具光滑的鳞片。菊水属只有一种菊水，原产墨西哥。

Whole plant flat; flowers covered with smooth squama; native to Mexico.

豪猪刺新绿柱
Stenocereus hystrix (Haw.) Buxb.

植株圆柱形，在基部分枝，呈灌木状，深黄绿色。具 12~16 条具节纵向直棱。刺座着生于节间，具 10~12 枚灰白色短周刺和 1~2 枚黄色中刺，长 2~2.5 cm。花侧生，白色。花苞为红褐色。原产墨西哥。

Whole plant terete, branched at base, shrub like, dark yellowish green; 12 to 16 straight ribs; flowers white; native to Mexico.

近卫柱
Stetsonia coryne (Salm-Dyck) Britton & Rose

为树形仙人掌，株高 5~8 米，下部株幅 40 cm，具 8~9 条棱，周刺 7~9 根，长 3 cm，中刺 1 根，长 8 cm，刺白色或黄褐色，随后渐变深至黑色。花白色。生长于温暖环境，原产南美玻利维亚。

Whole plant tree like, 5 to 8 m high, 40 cm wide at base, 8 to 9 ribs, spines white or yellow brown, gradually turn into dark black. Flowers white. Grow in warm condition; native to Bolivia, South America.

斯氏沟宝山
Sulcorebutia steinbachii (Werderm.) Backeb.

植株球形至长球形，高 6 cm，株幅 4 cm，生有约 13 条棱，周刺 6~8 根，黄色，褐色或近乎黑色，长 2.5 cm，花为浅红色、血红色或紫色。原产玻利维亚。

Whole plant spherical or long spherical; 6 cm tall, 4 cm wide, 13 ribs, 6 to 8 spines, spines 2.5 cm long, yellow, brown or nearly black; flowers pale red, blood red or purple; native to Bolivia.

白㭴仙人掌
Tacinaga inamoeua (K. Schum.) N.P. Taylor & Stuppy

植株为攀援类仙人掌，花生于茎基部，花管较长，花蕊突出，傍晚开，橙红色。原产巴西。

Whole plants climbing; flowers grow on the bottom of stems; flowers orange-red, bloom in the evening; native to Brazil.

大统领
Thelocactus bicolor (Galeotti ex Pfeiff.) Britton & Rose

花顶生，红紫。广布于美国得克萨斯州西南部和新墨西哥州。
Flowers reddish purple, at the apex. Widely distributed in southwestern Texas, and New Mexico, the United States.

多色玉
Thelocactus heterochromus (F.A.C. Weber) Osten

植株球形或长球形，株幅 15 cm，具 8~9 条贴近茎体的扁平棱，周刺 7~10 根，上层周刺向外扩散分布，中刺长 4 cm，全部刺为红色或褐色，花浅紫色，向内颜色较深。原产墨西哥。

Whole plant spherical or long spherical, 15 cm wide; 8 to 9 flat ribs; spines red or brown; flowers pale purple; native to Mexico.

天昊
Thelocactus hexaedrophorus (Lem.) Britton & Rose

株幅 15cm，具 12~13 条带有 6 个左右疣状突起的棱，周刺 6~9 根，长约 2cm，中刺 0~2 根，花白色或红色，原产墨西哥。

Whole plant 15 cm wide, 12 to 13 verrocose ribs; 6 to 9 outer spines, up to 2 cm long, 0 to 2 central spines; flowers white or red; native to Mexico.

图拉瘤玉球
Thelocactus tulensis (Poselger) Britton & Rose

花粉色、白色到黄色。分布于墨西哥塔毛利帕斯州和圣路易斯波托西州。

Flowers pink, white or yellow. Distributed in Tamaulipas and San Luis Potosi, Mexico.

月章
Trichocereus 'Variegata' (Hybrid)

植株为毛花柱属一个栽培黄金斑变种，由于植株缺少叶绿素故需要用量天尺 (Hylocereus undatus) 嫁接，植株柱形，颜色美观为较名贵仙人掌植物。

A colorful cultivar of genus Trichocereus; whole plants terete; a rare species.

劳氏陀螺果
Turbinicarpus laui Glass & R. A. Foster

球形，花为淡紫色生于植株顶端。分布于墨西哥圣路易斯波托西州。

Spherical, flowers purplish at the apex. Distributed in San Luis Potosi, Mexico.

精巧殿
Turbinicarpus Pseudopectinatus (Backeb.) Glass et Foster

株幅 3-6cm，疣突为斧形，具 40-56 根整齐的齿状白刺，花白色并具有紫红色条纹，原产墨西哥。

Whole plant 3 to 6 cm wide; 40 to 56 regular tooth like white spines; flowers white with purple red stripes; native to Mexico.

大花毛花柱
Turbinicarpus pseudopectinatus (Backeb.) Glass & R.A. Foster

植株高 35 cm，株幅 6 cm，具 14 条棱，周刺 8~9 条，中刺 1 条，花大血红色。原产阿根廷。
Whole plants 35 cm tall, 6 cm wide; 14 ribs; 8 to 9 outer spines, 1 central spine; flowers big, blood red; native to Argentina.

施氏升龙
Turbinicarpus schmiedickeanus subsp. schwarzii (Shurly) N.P. Taylor

植株球形近扁平,有1~2条弯曲软刺。花大,白色或奶油白色。原产墨西哥。
Whole plant spherical, 1 to 2 curving spines; flowers big, white or cream white; native to Mexico.

鲛丽球

Turbinicarpus subterraneus (Backeb.) A.D. Zimmerman

鲛丽球

Turbinicarpus subterraneus (Backeb.) A.D. Zimmerman

植株长球形，株高 10 cm，株幅 3~4 cm，周刺 16~25 根，白色，中刺 2~3 根黑色，刺座具白色绒毛，花顶生，较大，淡紫红色，花心黄色。原产墨西哥。

Whole plant long spherical; 10 cm tall, 3 to 4 cm wide; white tomentum grow on the areole; flowers big and purple-red, grow on top; native to Mexico.

瘤果鞭美玉恋
Weberocereus biolleyi (F.A.C. Weber) Br. et R.

植株长条形，具不规则的条棱，株幅 4~6 cm。高 80 cm，刺座生有 1~3 根非常小的刺，外花瓣深红色，内花瓣为粉白色。原产自墨西哥。

Whole plant multi-branched, 4 to 6 cm wide, 80 cm tall; 1 to 3 tiny spines grow on areole; outside petals dark red, inside petals pinkish white; native to Mexico.

卡氏花笠球
Weingartia kargliana Rausch

植株球形，具 10~12 条宽阔浅棱，刺座长有白色棉状绒毛，周刺 4~6 根白色，中刺 1~3 根，黑色。花顶生，较大，金黄色。原产玻利维亚。

Whole plant spherical; 10 to 12 broad ribs; white tomentum grow on 4 to 6 white outer spines, 1 to 3 black central spines; flowers big, golden, and acrogenous; native to Bolivia.

软毛花笠球
Weingartia lanata F. Ritter

植株长球形，刺座椭圆形，周刺 12~16 根，中刺最多为 15 根，黄色或黄褐色，花金黄色或淡黄色，顶生。原产玻利维亚。

Whole plant long spherical, areole oval; flowers golden or yellow, acrogenous; native to Bolivia.

威斯汀花笠球
Weingartia westii (Hutchison) Donald

植株长球形，刺座长在疣突上，并有白色棉状绒毛，周刺 10~16 根，白色，中刺 1~2 根黑色，茎顶长黑色中刺。花顶生，金黄色。原产玻利维亚。

Whole plant long spherical; white tomentum grow on areole; 10 to 16 outer spines, 1 to 2 black central spines; flowers golden and acrogenous; native to Bolivia.

红花威氏仙人掌
Wilcoxia viperina (F.A.C. Weber) Britton & Rose

植株灌木形，多分枝，茎深褐色，花红色，花蕊黄色，生于茎基部。栽培容易。原产美国南部和墨西哥等。

Whole plant shrub like; multi-branched; stems dark brown; flowers red, grow on the base of stems; easy to cultivate; native to southern U.S. and Mexico.

蟹爪兰
Zygocactus truncatus (Haw.) K.Schum.

为叶状仙人掌类，生有叶状扁平齿状茎，花生长在齿状叶茎顶，红色。原产巴西，现已广泛引种栽培到世界各地。

Stems flat and tooth like; flowers red, grow on top of stems; native to Brazil; widely introduced and cultivated.

安第斯山的仙人掌
Opuntia sp. (Andes, Peru)

安第斯山的仙人掌
Opuntia sp. (Andes, Peru)

安第斯山的仙人掌
Opuntia sp. (Andes, Peru)

安第斯山的仙人掌
Opuntia sp. (Andes, Peru)

安第斯山的凤梨科植物
Puya raimondii Harms (Bromeliaceae)

百合科 (Liliaceae)

百合科近期与其它科合并后成为一个庞大家族，约有 250 属 3700 多种，其肉质类植物主要分布在芦荟属 (Aloe)、十二卷属 (Haworthia) 和鲨鱼掌属 (Gasteria)，其中以芦荟属最为丰富，包括杂交栽培种超过了 500 种，很多种属由于形态、色泽多姿多彩，已成为家居庭院和植物公园的常见观赏花卉，芦荟属有些种类成为药物、保健品和化妆品的重要原料，被大量栽种。

After combining with other families, the Liliaceae become a large family, comprising about 250 genera and 3700 species. Succulent plants of this family are distributed mainly in the genera Aloe, Haworthia and Gasteria. There are more than 500 species, including hybrid cultivars, in genus Aloe, most of them are colorful and diversified in morphology. They become common ornamentals in private courtyards and plant parks. Some plants of genus Aloe are widely cultivated due to their medicinal and nourishing values.

具皮刺芦荟
Aloe aculeata 'Red'

叶片两面均着生小疣凸刺，叶片呈黄绿至橙黄色，看上去十分调和，为一种合适盆栽种类，原产非洲斯威士兰等地区。

Small warts and spines grow on both sides of the leaves; leaves yellowish green to orange; suitable for potting; native to Swaziland.

具皮刺芦荟锦
Aloe aculeata 'Variegata'

叶片背面分布很多小疣凸刺，而靠叶端呈橙红色，非常有特色。

Leaf apex orange-red; leaf back covered with little warts and spines; very particular.

疣突芦荟锦
Aloe aculeate 'Variegata'

为疣突芦荟一个具斑锦栽培变种。
A cultivar of Aloe aculeate.

具皮刺芦荟锦
Aloe aculeata 'Variegata' (Red)

植株呈莲座生长，叶背面生长很多小疣突，叶尖呈红色。
Whole plant grow in a rosette, leaf back verrucose, leaf apex red.

亚非尼加芦荟（非洲芦荟）与好望角芦荟杂交种
Aloe africana × A. ferox

阿非利加芦荟与马洛夫芦荟杂交种
Aloe africana × marlothii (Hybrid)

阿非利加芦荟
Aloe africana Mill.

茎细长，高约 2~4m。花序单生，每株可抽出数枝。花蕾向下，开花后上翘，橙黄色。分布于南非的南部杂树丛中。

Stems slender, 2 to 4 m high; inflorescence solitary, multi-branched; flower buds orange, grow downward, bend up after bloom; distribute in southern South Africa.

阿非利加芦荟
Aloe africana Mill.

相似芦荟
Aloe alooides (Bolus) Druten

植株高大，肉质叶长而向下弯，枯叶留在植株，显得更加宏伟，并增加其观赏价值，不少植物园均有引种。

Whole plant huge and tall; leaves long, succulent and decurved; withered leaves remain on the plants, which makes them more magnificent and increases their ornamental value; cultivated in many botanical gardens.

巴伯芦荟
Aloe barberiae T.-Dyer

高大乔木，可达 20m，多分枝。花为橙红色。分布于南非东至东南部沿海及莫桑比克、斯威士兰，喜生于稠密的矮林中。

Tall trees up to 20 m high; multi-branched; flowers orange-red; distribute in the east and southeast coast of South Africa, Mozambique and Swaziland; often grow in dense low forests.

巴特尔芦荟
Aloe barteri Baker

植株无茎，叶较宽、厚、肉质，密生呈莲座状排列，叶面具有明显的白色斑痕，叶缘具有褐色齿状刺，花序分枝，花淡红色，原产南部非洲。

Whole plant acaulescent; leaves broad, fleshy, and succulent, densely arrange in a rosette, white spots cover the leaf surface, brown tooth like spines grow on the leaf margin; inflorescence branched, flowers pale red; native to southern Africa.

巴哈那芦荟
Aloe berhana Reynolds

植株基本无茎。叶缘具棕色小齿状刺。花序较长，淡红色，花黄色，圆锥状。原产加勒比海。

Mostly acaulescent; brown dentoid spines grow on the leaf margin; inflorescence long, pale red; flowers yellow, conical; native to Caribbean Sea.

博伊尔芦荟
Aloe boylei Baker

植株基本无茎。花大，排列成头状，橙色，顶部带绿色。分布于南非东部。

Acaulescent; flowers big, orange, arrange in head shape; distribute in eastern South Africa.

短叶芦荟
Aloe brevifolia Mill.

一种较小型可盆栽的芦荟，肉质叶肥厚粗壮，叶尖带橙红色，花柄笔直，整个植株形态十分美观。

A relatively small potted Aloe; leaves succulent, fleshy and thick, leaf apex orange-red; flower stalk straight; particularly beautiful.

烛台芦荟
Aloe candelabrum Tod.

植株单茎，可高达2~5m。叶绿色转红色，向外弯曲，呈莲座排列。花鲜红色，长锥形，远看似一座点燃的巨型烛台，十分壮观。在南非是常见的庭院栽培芦荟种类。

A common courtyard potting species of genus Aloe in South Africa; single stem; 2 to 5 m high; leaf color from green to red, leaves arrange in a rosette and curve outward; flowers bright red, long conical, reminiscent of a huge burning candles.

红刺芦荟
Aloe brownii Baker

植株具有肥厚的肉质叶，呈莲座生长，有短壮棕红色刺生长在叶缘上。

Leaves succulent and fleshy, arrange in a rosette; short and brownish red spines grow on the leaf margin.

栗褐芦荟
Aloe castanea Schönland

植株群生。花长条，栗褐色，排成蛇形，弯曲。花蕊橙色，伸出花冠外。原产于南非北部。

Clustered; flowers chestnut, long and curving, arrange in S-shape; stamen orange, grow out of corolla; native to northern South Africa.

睫毛芦荟
Aloe ciliaris Haw.

植株攀援状。叶翠绿，较薄，螺旋向上生长。叶缘有睫毛状突起，特别是叶稍部分，故此得名。花一般红色，顶部常呈现橙红色。主要分布于南非南部的密集灌木丛中。

Leaves thin, emerald and spirally grow upward; leaf margin has eyelash shaped protuberance; flowers usually red, orange on the top; distribute in southern South Africa, grow in dense shrubs.

簇叶芦荟
Aloe comosa Marloth & A. Berger

茎高约 2m，花序比茎还高，故开花时可高达 5m，花序呈长纺锤形，花蕾粉红色。分布于南非西南部干旱地区。

Stems up to 2 m high, inflorescences are even higher than stems, whole plant up to 5 m high when blooming; inflorescence long and spindle-shaped; flower buds pink; distribute in the arid areas of southwestern South Africa.

康普顿芦荟
Aloe comptonii Reynolds

植株盆栽多单生，但露地栽培多为群生。花序梗分枝成头状花序，花深红至鲜红色。分布于南非南部。

Potted plants, often solitary; clustered when cultivate outdoor; peduncle branch into capitulum; flower color from dark red to bright red; distribute in southern South Africa.

圆锥芦荟
Aloe conifera H. Perrier

无茎。花序单生,长圆锥形,花黄色。分布于马达加斯加岛。

Acaulescent; inflorescence solitary, long conical; flowers yellow; distribute in Madagascar Island.

大肚芦荟
Aloe crassicaulis

其木质茎灰白肥大,这是整个植株的特点,其观赏价值颇高。原产南部和东部非洲。

Woody stem greyish white and fleshy; very particular; high ornamental value; native to southern and eastern Africa.

达市洛里斯芦荟
Aloe dabenorisana van Jaarsv.

匍匐茎分枝，形成多个莲座在一起倒挂在悬崖峭壁的石缝中。花序分枝，向上开花，花黄色带有红线。分布于南非西部极小范围的大山石英岩层的石缝中。

Stolon branched, arrange in multiple rosette, often suspend in the rock cracks of cliffs. Inflorescence branched, bloom upward; flowers yellow with red lines; grow in quartzite cracks in the limited areas of western South Africa.

隐柄芦荟
Aloe cryptopoda Baker

植株基本单生。花序圆锥形，花柄隐藏在苞片内；花有红、黄、黄红三色，盛开后转为黄色，十分美丽。

Mostly solitary; inflorescence conical; flower stalk hide in bract; flowers red, yellow or yellowish red and turn into yellow after full bloom.

威氏芦荟
Aloe dewetii Reynolds

植株单生，无茎。花序多分枝，高达 3m，花稀疏地排成圆柱形，红色。分布于南非东部和斯威士兰的开阔平缓草坡地上。

Whole plant solitary and acaulescent; inflorescence multi-branched, up to 3 m high; flowers red, sparsely arrange in column; distribute in the open, flat grass land of eastern South Africa and Swaziland.

远距芦荟
Aloe distans Haw.

茎沿地面爬行可长达 3m，多分枝，密集成丛。花序由许多红色的花密集组成。分布于南非西南沿海很小的范围内，喜海洋性气候。

Stem crawl along the ground, up to 3 m long; multi-branched and compactly clustered; inflorescence composed of red flowers; distribute in the southwest coast of South Africa; favor oceanic climate.

多花序芦荟
Aloe divaricata A. Berger

花序多分枝，可形成 20 个以上的长圆锥形，花红色。分布于马达加斯加岛西部。

Inflorescence multi-branched, could grow more than 20 long cones; flowers red; distribute in western Madagascar Island.

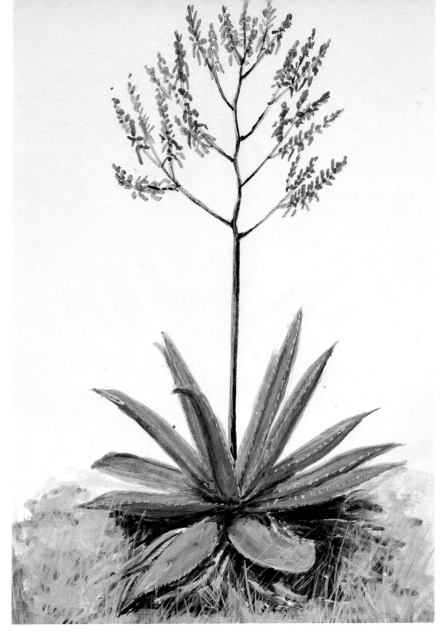

戴尔芦荟
Aloe dyeri Schönland

植株单生。花序多分枝，高达 2m，花深红色，顶部带灰色，基部膨大。分布于南非东北部高海拔山区地带荫蔽的河流峡谷和深谷中。

Solitary; inflorescence multi-branched, up to 2 m high; flowers dark red and grey on top, base huge and expanded; distributed in the high altitude mountain areas, valleys and canyons of northeastern South Africa.

短叶芦荟锦
Aloe echeveria Variegata

植株呈莲座生长，叶短，长三角形，肉质。花柄直立不分枝，橙红色为栽培种。为较理想的盆栽观叶花卉。

Whole plant arrange in a rosette; leaves succulent, short and long triangular. Flower stalk erect and non-branched; ideal potted ornamental.

高芦荟
Aloe excelsa A. Berger

植株高大宏伟，成年植株高达 4m。花深红色，排成直立柱状，远望像红火炬在燃烧，十分壮观。分布于南非东北角和津巴布韦、赞比亚、马拉维等地。

Whole plants huge and magnificent; mature plants up to 4 m high; flowers dark red, arrange in erect column, look like burning torches, very spectacular; distribute in northeastern South Africa, Zimbabwe, Zambia and Malawi.

好望角芦荟与马洛夫芦荟杂交种
Aloe ferox × marlothii (Hybrid)

好望角芦荟
Aloe ferox Mill.

花小，密集成圆柱状，花色多样有橙色、大红、黄色等。分布于南非东南沿海。

Flowers small, orange, red or yellow, densely arrange in column; distribute in the southeast coast of South Africa.

粉绿芦荟
Aloe glauca Mill.

植株无茎。花序梗粗壮且较高，直立，圆锥状。花橙红色。分布于南非南部以及刚果和纳米比亚的干旱地区。

Acaulescent; peduncle tall, thick, erect and conical; flowers orange; distribute in the arid areas of southern South Africa, Namibia and Congo.

球芽芦荟
Aloe globuligemma Pole-Evans

植株基本无茎。花柄较长，分枝，水平生长，单向排列。花芽球状，红色，盛开时变为粉红至白色。花短，顶部膨起。这是该种芦荟的显著特征。分布于津巴布韦、南非北部，多见于温暖稀树草原。

Basically acaulescent; flower stalk long and branched, horizontally grow in the same direction; flowers red and bud ball like, turn into pink or white during full bloom; flowers short, top huge and expanded; distribute in northern South Africa and Zimbabwe; often seen in the warm and treeless grass land.

哈迪芦荟
Aloe hardyi Glen

在茎基部多分枝，倒挂在峭壁的石缝中。花序单生向下，短圆锥形，花红色。分布于南非北部偏东山区的一小范围，生于峭壁的岩石缝中。

Multi-branched at the bottom of stems; suspend in the rock cracks of cliffs; short conical inflorescence solitary, grow downward; flowers red; distribute in limited areas of northern South Africa, grow in the rock cracks of cliffs.

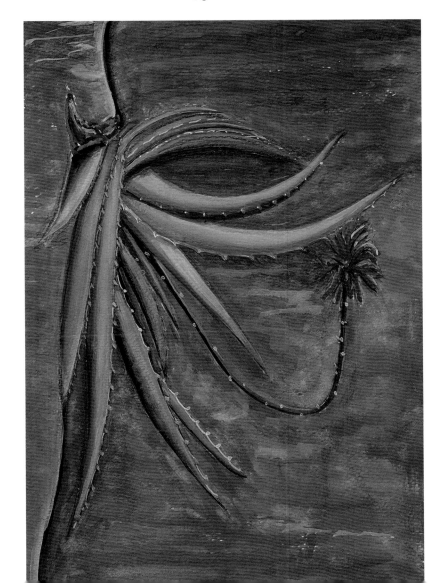

艳丽芦荟
Aloe hexapetala Salm-Dyck (Aloe speciosa Baker)

亦称歪头芦荟，前名是由于其整个花序非常艳丽，而且硕大，后者名则是植株顶端是歪斜的。艳丽芦荟生长粗犷，体型高大美观，为不少植物园所引种的好选择。

Inflorescence huge and gorgeous; top of the plants askew; easy to cultivate, introduced by many botanical gardens due to their huge and beautiful morphology.

伊比戴芦荟
Aloe ibitiensis H. Perrier

花序多分枝，粉红或红色。分布于马达加斯加岛。
Inflorescence multi-branched, pink or red; distribute in Madagascar Island.

喀来斯芦荟（老树）
Aloe khamiesensis Pillans

茎直立高，一般不分枝，叶斜向上伸展，上下面均有白色小斑点，花梗分枝，花红色，盛开后变黄，成为美丽的双色花，原产南部非洲的内陆高原。
Stems erect and tall, usually unbranched; leaves covered with white spots, grow upward; pedicel branched, flowers red, turn into yellow after bloom; native to the inland plateau of southern Africa.

喀来斯芦荟
Aloe khamiesensis Pillans

花梗分枝成数个长圆锥形。花红色，盛开后渐变黄色，成双色花。分布于南非西部内陆高原的大山中。

Pedicel branch into multiple long cones; flowers red, turn into yellow after full bloom; distribute in mountains of western South Africa inland plateau.

红线芦荟
Aloe lineata (Aiton) Haw.

植株不高，其特点是花蕾硕大，为花苞所包裹，苞片苞在花柄上，粉红色至红色，颇具观赏价值。

Whole plant short and featured by its huge flower bud, flower stalk is covered by bracts, flowers pink to red, high ornamental value.

海滨芦荟
Aloe littoralis Baker

植株较高大。花粉红至红色，有时盛开后渐转为黄色。分布于南非北部以及博茨瓦拉、安哥拉、纳米比亚、津巴布韦和莫桑比克等地。

Whole plants tall; flowers pink to red, sometimes turn into yellow after full bloom; distribute in northern South Africa, Botswana, Angola, Namibia, Zimbabwe and Mozambique.

长苞芦荟
Aloe longibracteata Pole-Evans

植株矮小。花梗较高直立，有分枝，花红色，较稀疏。分布于南非。

Whole plant short; pedicel tall, erect and branched; flowers red and sparse; distribute in South Africa.

变黄色芦荟
Aloe lutescens Groenew.

花尖塔形，顶端鲜红，花基部为亮黄色，十分显眼。分布于南非和斯威士兰。
Flower Spires tower-shaped, top bright red, bottom bright yellow, very particular; distribute in South Africa and Swaziland.

水晶芦荟
Aloe macroclada Baker

为一种芦荟。
A soecies of Aloe.

斑痕芦荟
Aloe maculata All.

无茎。花序柄粗壮，分枝，花密集而短，红色，也有橙色和黄色。广泛分布于南非东部和南部、津巴布韦。喜温暖的海洋性气候。

Acaulescent; inflorescence stem thick, solid and branched; flowers dense and short, often red, sometimes orange or yellow; widely distribute in southern and eastern South Africa and Zimbabwe; favor warm oceanic climate.

皂素芦荟
Aloe maculata All.[Aloe saponaria (Aiton) Haw.]

植株叶缘有明显褐色齿，叶有斑点，叶汁液在水中能产生肥皂泡沫。原产南部非洲。

Leaves spotted; the water solution of leaf juice could generate soap bubbles; native to southern Africa.

马洛夫芦荟与好望角荟杂交种
Aloe marlothii × ferox (Hybrid)

马洛夫芦荟

Aloe marlothii A. Berger

植株直立高大。叶缘具明显的齿状刺，叶背亦具刺点。花序梗水平向上或斜向上分枝。花单边着生，排列成牙刷状，橙红色和黄色的均见。喜欢生长在开阔矮灌木稀树草原和岩石露头处，人工栽培较容易，常见于南部非洲公园和庭院中。分布于南非东北部，向西至莫桑比克，向北至博兹瓦纳和津巴布韦。

Whole plants huge and erect; dentoid spines grow on the leaf margin, spines also grow on the leaf back; peduncle branch upward; flowers only grow on one side, arrange in toothbrush shape; flowers orange-red or yellow; often grow on wide, low shrubs grass land; easy to cultivate; often seen in the gardens and courtyards of southern Africa; distribute in northeastern South Africa, Mozambique, Botswana and Zimbabwe.

马洛夫芦荟

Aloe marlothii A. Berger

马洛夫芦荟

Aloe marlothii A. Berger

马洛夫芦荟

Aloe marlothii A. Berger

黑刺芦荟

Aloe melanacatha A. Berger

植株初生为单生，成长后在基部萌生出多个新芽变成丛生。花序单生，长圆锥形。花蕾红色，开放后转为橙黄色。分布于南非西部和纳米比亚南部干旱贫瘠的石砾地上。

Solitary when newborn, sprout at the bottom after mature; inflorescence solitary, long conical; flower buds red, turn into orange after bloom; distribute in the arid areas of western southern Africa and south Namibia.

易变芦荟
Aloe mutabilis Pillans

植株茎矮短，基部少量分枝，直立或蔓生。花蕾红色，开花后渐转为黄色，花序有时呈双色。分布于南非东北部高原的悬崖石缝中。

Stems short, branched at the bottom; erect or sprawl; flower buds red, turn into yellow after bloom; inflorescence sometimes bicolor; grow in cliff cracks, distribute in the plateau of northeastern South Africa.

易变芦荟
Aloe mutabilis Pillans

佩格勒芦荟
Aloe peglerae Schönland

多为单生，叶内弯成球形。花序单生，花蕾鲜红色，开花后转黄，深紫色花蕊伸出花外。分布于南非北部内陆一小范围的岩石山北坡。

Mostly solitary; leaves curve into sphere; inflorescence solitary, flower bud bright red, turn into yellow after bloom; stamen grow out flowers; distribute in the inland areas of northern South Africa, grow on the north slope of rocky mountain.

希帕尔芦荟
Aloe percrassa Tod. (Aloe schimperi Tod.)

希帕尔芦荟
Aloe percrassa Tod. (Aloe schimperi Tod.)

植株无茎。花序梗分枝，花红色排成圆锥状。分布于东部非洲、马达加斯加岛。

Acaulescent; Peduncle branched; flowers red, conically arranged; distribute in eastern Africa and Madagascar Island.

巨箭筒芦荟
Aloe pillansii L. Guthrie

高大乔木。花序多分枝,花黄色。分布于南非西部及纳米比亚。为南非著名的特色植物。

Distinctive plants of South Africa; tall trees; inflorescence multi-branched; flowers yellow; distribute in western South Africa and Namibia.

巨箭筒芦荟(比兰斯芦荟)
Aloe pillansii L. Guthrie

扇状芦荟
Aloe plicatilis (L.) Mill.

植株高大呈乔木状。茎从基部起多分枝，叶对生，成两排像打开的折扇。花红色。分布于南非最南端。

Whole plants huge and tree like; stems branched from the bottom; leaves opposite, look like an open fan; flowers red; distribute in the southernmost of South Africa.

多齿芦荟与红线荟杂交种
Aloe pluridens × lineata (Hybrid)

其笔直花柄，鲜红花序和宿留枯叶使其深爱栽种者所喜欢。

Flower stalk straight; inflorescence bright red; deeply favored by plant enthusiasts.

多齿芦荟
Aloe pluridens Haw.

亦称法芝西芦荟，植株直立，叶细长向下垂，有点像穿上"唐装上衫的书生"。

Whole plants erect; leaves slender, grow downward; look like a scholar wearing a Tang suit.

多叶芦荟
Aloe polyphylla Pillans

植株多单生。叶片三角形螺旋状排列成圆盆状。花红色，花柄有分枝，并有白色苞片。分布于南非海拔2000m潮湿高原草地中，是南非最具特色的芦荟之一。

One of the most distinctive Aloe in South Africa; usually solitary; leaves triangular, spirally arrange in a basin shape; flowers red, flower stalk branched with white bract; distribute in the 2000 m altitude humid grass highland of South Africa.

比勒陀尼亚芦荟
Aloe pretoriensis Pole-Evans

叶密集成一个大莲座，直立。蓝绿色。花序高达 4~5m，花红色，偶见橙红色。分布于南非北部、斯威士兰和津巴布韦的岩石缝中。

Leaves blue-green, erect, densely arrange in a big rosette; inflorescence 4 to 5 m high, flowers red, rarely orange red; grow in the rock cracks of northern Africa, Swaziland, Zimbabwe.

赖茨芦荟
Aloe reitzii Reynolds

单生。花序成熟时分枝，花外面深红色，向花序柄一面黄色，每朵花有两色。分布于南非东北部小范围的草原石头坡上。

Solitary; inflorescence branch when they are mature; flowers bicolor; distribute in limited areas of northeastern South Africa, grow on rocky slopes of grass land.

赖茨芦荟
Aloe reitzii

植株丛生。花序多单生，花红色，花苞较大。分布于南非开普敦西南山区。

Clustered; inflorescence solitary; flowers red, flower bud big; distribute in the mountains of southwestern Cape Town, South Africa.

喜石芦荟

Aloe rupestris Baker

植株在基部多分枝。花序单生,直立,筒状,花色有红色、橙色和黄色等。分布于南非东部和斯威士兰、莫桑比克南部,喜生于高温的峡谷。

Multi-branched at the bottom; inflorescence solitary, erect, and columnar; flowers red, sometimes orange or yellow; distribute in eastern South Africa, Swaziland, southern Mozambique, usually grow in warm valleys.

苏拿利芦荟
Aloe somaliensis C.H. Wright ex W.Watson

植株矮小。叶棕褐色，有较多横向条状白色斑痕，常见会卷曲。为一种较少见的芦荟品种。分布于南非局部地区。

A rare species of genus Aloe; whole plants short; leaves brown with horizontal white stripe, often curly; distribute in certain areas of South Africa.

索潘斯伯格芦荟
Aloe soutpansbergensis Verd.

丛生。花序单生，花大，橙红色至红色。分布于南非北部高海拔多雨的山坡石缝中。

Clustered; inflorescence solitary; flowers big, from orange-red to red; distribute in northern South Africa, grow in the rock cracks of high altitude rainy slopes.

奇丽芦荟与开普敦芦荟杂交种
Aloe speciosa × A. ferox

穗花芦荟
Aloe spicata L. f.

叶缘具密刺，花穗状，原产南部非洲，莫桑比克和斯威士兰。
Leaf margin covered with dense spines; flowers spiciform; native to southern Africa, Mozambique and Swaziland.

珊瑚芦荟
Aloe striata Haw.

植株无茎。花梗多分枝，花聚集于顶部，花有粉红色、鲜红色、橙红色和黄色等。分布于南非南部和西部，以及纳米比亚南部干旱地区。

Acaulescent; peduncle multi-branched; flowers pink, bright red, orange-red or yellow, gather on top; distribute in southern and western South Africa and the arid areas of southern Namibia.

多浆芦荟
Aloe succotrina Lam.

植株较大。花序分枝，成圆筒状，橙红色。分布于南非开普敦。

Whole plants big; inflorescence orange, branched and columnar; distribute in Cape Town, South Africa.

开卷芦荟

Aloe suprafoliata Pole-Evans

开卷芦荟
Aloe suprafoliata Pole-Evans

在盆栽环境下，植株叶片呈对称向下弯曲，像打开一本书籍似的，故此得名，但在野外则有些不同。

In potted condition, leaves symmetrically curve downward, like an open book, but the wild types are different.

拉思卡芦荟
Aloe thraskii Baker

植株高大，最高可达 8m。花序多分枝呈短圆柱状。花橙色，花蕊为鲜红色，伸出花冠外。分布于南非东部、斯威士兰和莫桑比克，喜生长于温暖山谷和树丛中。

Whole plants huge and tall, up to 8 m high; inflorescence multi-branched and short columnar; flowers orange, stamen bright red, grow out of corolla; distribute in eastern South Africa, Swaziland and Mozambique. Favor to grow in warm valleys.

沙丘芦荟
Aloe thraskii Baker

茎粗壮，高约 2m 以上。花序分枝可达 20 多支，短圆柱形，花黄色，橙色花蕊伸出花外。分布于南非东部沿海植被茂密的灌木覆盖的沙丘上。

Stems thick and solid, up to 2 m high; inflorescence branched, up to 20 branches; short columnar; flowers yellow, stamen orange and grow out of flowers; distribute in the east coast of South Africa, grow on the shrub covered dunes.

钦科罗斯芦荟
Aloe unfoloziensis Reynolds

植株单生，无茎。花序粗壮直立。可高达1.5m，顶部有分枝，花红色，基部膨大。分布于南非东部河流峡谷中的稀树杂草丛中。

Solitary, acaulescent; inflorescence thick, solid and erect; up to 1.5 m high, branched on top; flowers red; bottom huge and expanded; distribute in the valleys of eastern South Africa, grow in treeless grass.

斑叶芦荟
Aloe variegata L.

植株高25cm，无茎。花序分枝，花大，红色。分布于南非西南部干旱地区。

Whole plants 25 cm tall; acaulescent; inflorescence branched; flowers big and red; distribute in the arid areas of southwestern South Africa.

瓦奥姆比芦荟
Aloe vaombe Decorse & Poiss.

植株高大。花梗多分枝，花红色排成圆锥状。分布于马达加斯加岛。

Whole plants huge and tall; peduncle multi-branched; flowers red, arrange in conical; distribute in Madagascar Island.

福尔肯芦荟与好望角芦荟杂交种
Aloe volkensii × ferox (Hybrid)

福尔肯芦荟与好望角芦荟杂交种
Aloe volkensii × ferox (Hybrid)

福尔肯芦荟
Aloe volkensii Engl.

植株高大直立，可达 4~5m。花梗分枝，花红色，圆筒形。分布于坦桑尼亚。

Whole plants huge, tall and erect, 4 to 5 m high; peduncle branched, flowers red, columnar; distribute in Tanzania.

费雷赫德芦荟
Aloe vryheidensis Groenew.

植株单茎。花序单生不分枝，花蕾红褐色，盛开时黄色。分布于南非北部山区。

Stems solitary; inflorescence solitary with no branches; flower bud red brown, turn into yellow when blooming; distribute in the mountain areas of northern South Africa.

费雷赫德芦荟
Aloe vryheidensis Groenew.

威肯斯芦荟
Aloe wickensii Pole-Evans

植株无明显茎，叶黄绿色，直立向上伸展，叶边缘具刺，花序分枝或单生，排成圆锥形，花黄色，原产非洲南部。

Whole plant acaulescent; leaves yellowish green, erect, spines grow on the leaf margin; inflorescence branched or solitary, arrange in a cone, flower yellow; native to southern Africa.

具皮刺芦荟杂交种
Aloe 'Hybrid'

薄叶芦荟

Aloe sp.

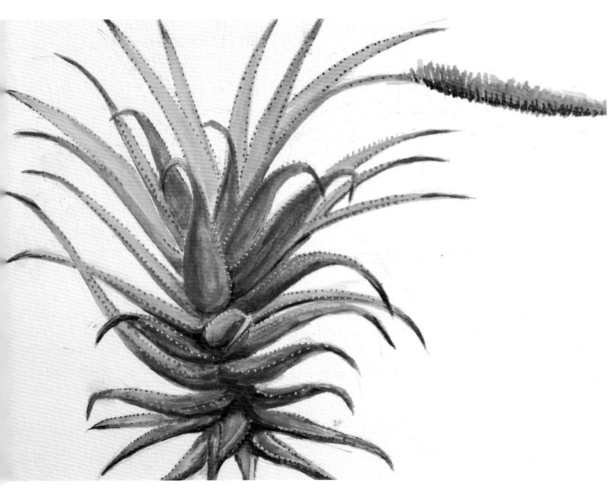

未鉴定芦荟一种

Aloe sp.

重叶芦荟

Aloe sp.

植株翠绿色,叶重叠生长,呈塔形,为栽培杂交种,是较理想盆栽观叶花卉。

A cultivated hybrid; whole plants emerald; leaves overlap with each other, tower like; ideal potted ornamental.

卧牛
Gasteria carinata (Mill.) Duval

卧牛

Gasteria carinata (Mill.) Duval
[Gasteria excavate (Willd.) Haw.]

植株矮小，叶翠绿色肥厚，有白色环痕，对生。易栽种，是一种常见盆栽小型观叶花卉。

A common small potted ornamental; whole plants short and small; leaves opposite, emerald and fleshy, with white circular marks; easy to cultivate.

纯叶牛脷锦
Gasteria carinata 'Variegata'

植株小，叶对生，成长后叶呈顺时针扭转，叶肥厚长条状，叶橙黄色，叶面有白色斑痕，为较名贵盆栽观叶花卉，是栽培变种。

A cultivated hybrid, a rare potted ornamental; whole plants small; leaves opposite, fleshy and long; leaves orange with white marks.

青龙刀
Gasteria disticha (L.) Haw.

植株体态是否像把"青龙刀"，那见仁见智。

Whether or not it looks like a dragon blade, different people, different views.

阔叶青龙刀
Gasteria disticha (L.) Haw. (Hybrid)

为青龙刀一个栽培变种。
A cultivar of Gasteria disticha.

阔叶青龙刀锦
Gasteria disticha 'Vartegata'

为青龙刀阔叶变种一个斑痕变种。
A variant of Gasteria disticha var. Hybrid.

短叶牛舳
Gasteria excavate 'Hybrid'

斑叶牛利
Gasteria sp.

为一较常见沙鱼掌属盆栽观叶花卉，生长较粗犷，容易栽种。
A common ornamental of genus Gasteria; easy to cultivate.

大叶青龙刀
Gasteria sp.

植株叶肥大,叶面黄绿,叶背翠绿色,具横向较整齐白色斑痕,叶对生,是一种非常常见盆栽观叶花卉。为一栽培杂交种。

A cultivated hybrid, very common potted ornamental; leaves opposite and fleshy, leaf surface yellowish green, leaf back emerald with regular white marks.

金城
Haworthia attenuata (Haw.)Haw.

多年生草本。叶片黄绿色,三角形。植株艳丽,是观赏价值颇高的观叶花卉。

Perennial herb; leaves yellowish green, triangular; whole plants bright colored, high ornamental values.

水晶掌
Haworthia attenuata var. radula (Jacq.) M. B. Bayer
[Haworthia radula (Jacq.) Haw.]

为百合科十二卷属一个栽培种。叶色淡绿色呈半透明，为一理想盆栽观叶花卉。

A cultivar of genus Haworthia; leaves pale green and semitransparent; an ideal potted ornamental.

雅致十二卷
Haworthia fasciata 'Elegans'

为十二卷属的一个栽培品种，亦被称作雉鸡尾，的确有点似雄性雉鸡尾毛。
A cultivar of genus Haworthia.

琉璃殿锦
Haworthia limifolia Marloth

为琉璃殿一个栽培变种，其体态和叶型叶色颇为栽种者喜爱。

A cultivar of Haworthia limifolia; deeply favored by plants enthusiasts.

霜鹤
Haworthia herbacea 'Variegata'

为百合科十二卷属一个栽培品种。
A cultivar of genus Haworthia.

瑞鹤

Haworthia marginata (Lam.) Stearn

为一较大型盆栽观叶花卉，肉质叶较坚实，原产南部非洲和西南非洲，现已被广泛引种各地区。

A relatively huge potted ornamental; leaves succulent and solid; native to southern and southwestern Africa; widely introduced all around the world.

霜百合

Haworthia reinwardtii (Salm-Dyck) Haw.

老株会匍匐，叶三角形呈螺旋密集排列生长，叶硬，背面生有10~12行排列整齐的小疣点，可分株繁殖。原产自南部非洲。

Leaves triangular, spirally arrange; 10 to 12 lines of little warts cover the leaf back; propagate by suckering; native to southern Africa.

翡翠十二卷

Haworthia sp.

为百合科十二卷属一个未定名的栽培种。
An unnamed cultivar of genus Haworthia.

五十之塔锦

Haworthia viscosa 'Variegata'

虽名为"锦",但未见颜色出现。
They are not very colorful, although they have been called 'Jin'.

景天科 (Crassulaceae)

景天科约有 30 多个属 1500 余种，为多年生低矮灌木，有些为藤本，是肉质植物重要的科属之一。分布较为广泛，在温暖、干燥地区都能看见它们。叶对生、互生和或轮生，绝大部分肉质化，为此其生长形态多样，色泽变化鲜艳，为花卉爱好者所喜欢，现已被广泛引种到世界各地，是盆栽观叶植物的重要成员。

The plants of family Crassulaceae, include approximately 1500 species and 30 genera, are generally perennial low shrubs but sometimes vine. They are one of the major family of succulent plants, and are extensively distributed in both warm and/or dry areas. Their leaves are opposite, alternate or verticillate, and mostly succulent. Due to the diversity of their growth morphology and bright color, the plants of family Crassulaceae are deeply favored by flower and plant enthusiasts and widely introduced all around the world, which makes them an important member of potted ornamentals.

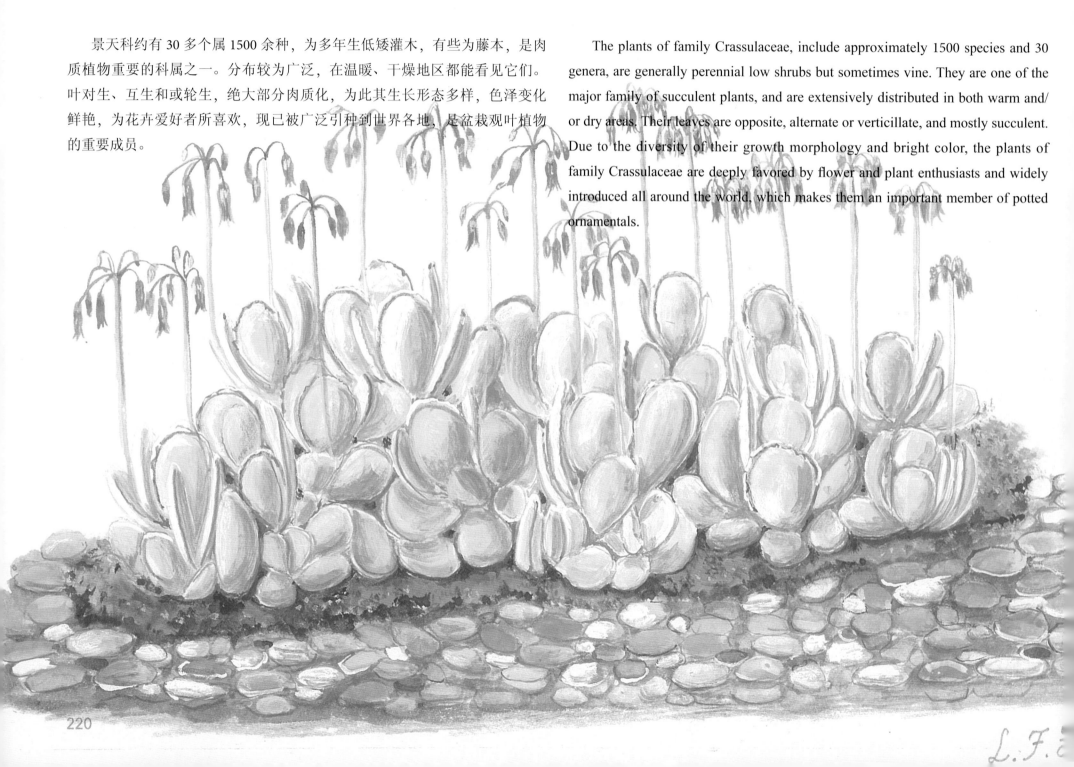

天章
Adromischus cristatus (Haw.) Lem.

为天锦章属多年生肉质草本观叶花卉，叶厚，呈波浪状，灰绿色，十分优雅，很适合盆栽。

Belongs to the genus Adromischus; perennial herbaceous succulent ornamentals; suitable for potting. Their elegant greyish-green leaves are thick and wavelike.

碎叶天章
Adromischus sp.

为一栽培种。

A cultivar of Adromischus.

筒形百合莲花掌
Aeonium (Hybrid)

植株筒形，叶贝壳状螺旋生长，叶根翠绿，叶中黄绿，叶缘枣红色，形态特别颇受栽培者喜爱。

Whole plant cylindrical; leaves shell like and grow spirally; leaf root emerald, middle yellowish green, leaf margin claret; deeply favored by plant enthusiasts.

包叶莲花掌
Aeonium (Hybrid)

为一栽培种。

A cultivar of Aeonium.

长叶莲花掌

Aeonium arboreum (Hybrid)

卷叶莲花掌

Aeonium arbereum (Hybrid)

为一栽培种。

A cultivar of Aeonium arberuum.

扭叶莲花掌
Aeonium arboreum (Hybrid)

为一栽培种。
A cultivar of Aeonium arberuum.

黑法师
Aeonium arboreum 'Atropurpureum' (Hybrid)

为莲花掌一个变种,亚灌木,高可达1米以上,整个形态有点像穿了黑大衣的法师。
A variant of Aeonium arboretum; subshrubs up to 1 m tall. It looks like a mage wearing a black robe.

黑法师

Aeonium arboreum 'Atropurpureum'(Hybrid)

为一栽培种。

A cultivar of Aeonium arbereum.

黑法师

Aeonium arboreum 'Atropurpureum'(Hybrid)

为一栽培种。

A cultivar of Aeonium arbereum.

黑法师
Aeonium arboreum 'Atropurpureum'(Hybrid)

为一栽培种。
A cultivar of Aeonium arbereum.

垂叶黑法师
Aeonium arboreum 'Atroprupureum'(Hybrid)

黑法师

Aeonium arboreum 'Atropurpureum'(Hybrid)

为一栽培种。

A cultivar of Aeonium arbereum.

黑法师

Aeonium arboreum 'Atropurpureum'(Hybrid)

为一栽培种。

A cultivar of Aeonium arbereum.

尖叶黑法师
Aeonium arboreum 'Hybrid'

红叶黑法师
Aeonium arboreum 'Hybrid'

为一栽培种。
A cultivar of Aeonium arbereum.

黑法师

Aeonium arboreum 'Atropurpureum'(Hybrid)

为一栽培种。

A cultivar of Aeonium arbereum.

山地玫瑰
Aeonium 'Greenovia diplocgcla'

肉质叶呈莲座状排列，叶色灰绿或翠绿色。原产于加那利群岛。由于造型美观别合适小型盆栽，现已成为世界上著名盆栽观叶花卉。
Leaves succulent, greyish green or emerald and arrange in a rosette; native to Canary Islands; particularly beautiful and suitable for potting; become a famous potted foliage plants worldwide.

山地玫瑰

Aeonium diplocyclum (Webb ex Bolle) T.H.M. Mes
(Greenovia diplocycla Webb ex Bolle)

红边莲花掌
Aeonium haworthii Webb & Berthel.

肉质叶较薄，呈莲座生长，叶缘为红色故此得名。
Leaves (with a red leaf margin) thin and grow in a rosette.

明镜
Aeonium tabulitorme (Haw.) Webb & Berthel.

叶较宽阔，紧密呈莲座生长，有点像绿色的镜。
Leaves broad and closely arrange in a rosette; looks like a green mirror.

莲花掌
Aeonium Webb & Berthel.

灌木或亚灌木，原产北非较为干旱的地方和非洲加纳利群岛，易栽培，现世界各地均有引种栽培，并培育出多个变种和栽培种。

Shrubs or subshrubs; native to the arid areas of northern Africa and Canary Islands; easy to be cultivated and widely introduced all around the world. Several variants and cultispecies of Aeonium arboretum are also been cultivated.

莲花掌
Aeonium Webb & Berthel.

皱叶红覆轮

Cotyledon macrantha A. Berger (Hybrid)

为红覆轮一个栽培品种。

A cultivar of Cotyledon macrantha.

红覆轮锦

Coryledon macrantha 'Red Variegata'

为银波锦一个皱叶栽培变种。

A cultivar of Cotyledon undulate; leaves wrinkled.

波浪叶银波锦
Cotyledon undulata 'Hybrid'

是银波锦具波浪形皱叶栽培品种。
A cultivar of Cotyledon undulata; leaves wrinkled and wavelike.

红边银波锦
Cotyledon undulata Haw.

多年生草本，植株可高达 40~50cm。叶片大、扇形、肉质翠绿色，叶缘红色，直立。是一种颇受欢迎的盆栽观叶花卉。
Perennial herb, 40 to 50 cm high; leaves big, fanshaped, succulent, and emerald, leaf margin red and erect; deeply favored by plant enthusiasts.

银波锦
Cotyledon undulata Haw.

原产非洲安哥拉，纳米比亚和南非，为常绿亚灌木，个头颇高大，但也适合盆栽，为常见的盆栽观叶花卉，观赏价值较高。

Evergreen tall subshrubs; native to Angola, Namibia, and South Africa. They are common potted ornamental with high value.

紫叶银波锦
Cotyledon undulata Haw.

为银波锦一个紫色叶栽培品种。各种鲜艳颜色的肉质叶，都十分受人喜爱，但大都缺少叶绿素，栽种不是很容易的。

A cultivar of Cotyledon undulata with purple leaves. Their bright-colored succulent leaves are deeply favored. However, they are relatively hard to cultivate due to the lack of chlorophyll in the leaves.

火焰银波锦
Cotyledon undulata var. haematophyllas

为红边银波锦一个艳红色的栽培品种，十分为人喜爱。
A favored cultivar of Cotyledon undulata with bright-red colored leaves.

筒叶青锁龙
Crassula 'Gollum'

植株分枝较多，形似小长喇叭，喇叭型分枝茎口有红色环，十分抢眼，生长较粗犷。
Usually multi-branched; trumpet like; easy to cultivate.

星都
Crassula 'Morgan's Beauty'

植株翠绿色，由肉质阔厚叶组成塔形莲座，非常美观特殊。可能是个青锁龙属的栽培种。

A cultivar from genus Crassula; Whole plants emerald; leaves broad, succulent and fleshy, arrange in tower like rosette, very particular and beautiful.

红花月
Crassula arborescens 'Haematophyllas'

植物易成群生，叶色随生长时间变长由绿转橙色再转变成橙红色至鲜红色。

Usually clustered. The color of their leaves changes over time from green to orange, then to orange-red, and finally to bright-red.

红花月

Crassula arborescens (Mill.) Willd.

玉春

Crassula barklyi N. E. Br

植株呈塔形，由不规则浅褐色肉质叶组成，非常特别。
Whole plants tower like; leaves succulent and pale brown, very particular.

火祭
Crassula capitella 'Flammeus'

为头状青锁龙一非常艳丽的栽培品种，火祭一名可能来源日本，其叶红似火之意吧。

A gorgeous colorful cultivar of Crassula capitlla. The name Crassula Canitella 'Flammeus' may originated from Japan, due to the resemblance between their leaves and burning fire.

红背青锁龙
Crassula capitella 'Red'

亦称尖红青锁龙，为头状青锁龙一个变种，原产非洲东南部。

A variant of Crassula capitella; native to the southeastern Africa.

头状青锁龙
Crassula capitella Thunb.

为较常见青锁龙属盆栽品种，原产非洲东南部。
A common cultivar of Crassula lycopodioides; native to the southeastern Africa.

塔形青锁龙
Crassula deceptor Schönland & Baker F.

亦称锥形青锁龙，植株呈锥形或尖塔形生长，取材于加拿大西海岸露地栽种，常见为群生。
Originate from an outdoor cultivar cultivated in the west coast of Canada. Usually clustered; look like cones or steeples.

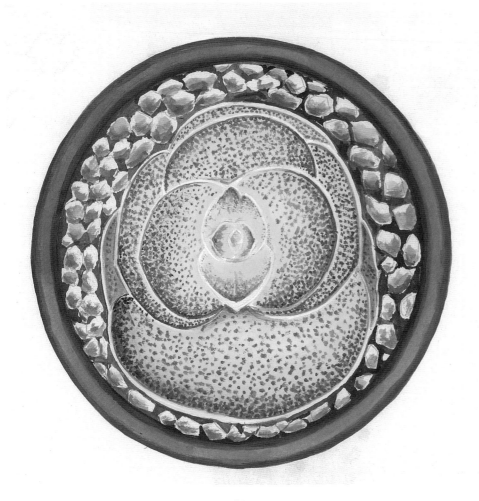

彩巴
Crassula hemisphaerica Thunb. (Hybrid)

植株黄绿色并长有紫色斑点，由大小扁平肉质叶组成莲座生长，十分美观。是青锁龙属一个栽培变种。

A cultivar from genus Crassula; whole plants yellowish green with purple spots; leaves flat and succulent, arrange in rosette, very beautiful.

神刀
Crassula perfoliata var. falcata (J. C. Wendl.) Toelken. (Crassula falcata J. C. Wendl.) (Hybrid)

为青锁龙属一个栽培种，生长粗犷，常见露地栽种，原产非洲南部。

A cultivar from genus Crassula; usually cultivated outdoor; native to the southern Africa.

神刀
Crassula perfoliata var. falcata (J. C. Wendl.) Toelken.
(Crassula falcata J. C. Wendl.) (Hybrid)

乔木状青锁龙
Crassula sp.

植株形似高大乔木，由灰绿色对生叶组成，很特别。是青锁龙属的一个栽培变种。

A cultivar from genus Crassula; tree like; leaves greyish green, opposite, very particular.

圆叶玉春
Crassula sp.

植株白色并长有翠绿色斑点，是玉春的一个栽培变种。

A cultivar of Crassula barklyi N.E.Br; whole plants white with emerald spots.

圆叶状青锁龙
Crassula sp.

植株灌木状，分枝，叶绿色倒卵行形，花顶生，白色。为青锁龙属一个栽培变种。

A cultivar from genus Crassula; shrub like; branched; leaves green and obovate; flowers acrogenous and white.

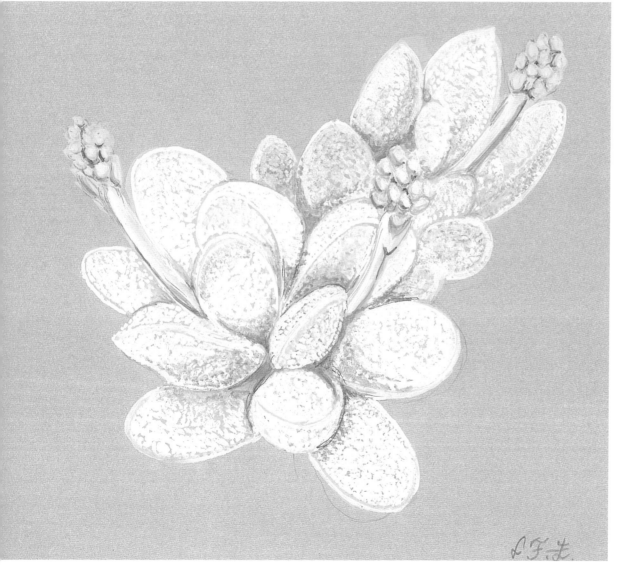

竹节青锁龙
Crassula sp.

植株灌木状，翠绿色轮生于茎节上，花顶生，红色，花心白色。生长粗犷，易栽培。

Shrub like; emerald verticillate and grow on stem nodes; flower petals red, center white, acrogenous; easy to cultivate.

白粉草
Dudleya pulverulenta (Nutt.) Britton & Rose

亦称粉叶草，其叶面着生密密细微白色绒毛，似一层白粉涂在叶面上，故此得名。取材于墨西哥北部海边渡假地，拍摄"冰海沉船"餐厅外景处。

Also called 'Fen Ye Cao' in Chinese because there are a layer of fine white tomentum, which looks like a layer of white powder, tightly covers the surface of their leaves. Originated from a beach resort in northern Mexico, where the restaurant scene of 'A Night to Remember' were filmed.

雪莲
Echereria laui Moran & J. Meyrán

多年生肉质草本，株高10~15cm，叶圆匙形，叶面布满白色粉状绒毛，花橙色，花期初夏至秋季，为常见盆栽种类。

A common cultivated species; perennial succulent herb, 10 to 15 cm high; leaves round spoon shaped, covered with white powder like tomentum; flowers orange, florescence from early summer to autumn.

紫红阔叶石莲
Echeveria (Hybrid)

为一栽培种。

A cultivar of Echeberia.

黄玉石莲花

Echeveria 'Yallum' (Hybrid)

可能是阔叶石莲花一个栽培品种。

This cultivar is possibly originated from Echevria eurychlamys.

艳红石莲（东云红叶）

Echeveria agavoides (Hybrid) Lem.

吉娃莲
Echeveria chihuahuaensis Poelln.

植物多为单生，是一种较为常见盆栽品种，原产南美洲。
Usually solitary, a common potted species; native to South America.

优雅石莲花
Echeveria elegans Rose

原产美洲，现已广泛被引种世界各地，作为较为常见盆栽观叶花卉。
A common potted ornamental; native to America and widely introduced all around the world.

小囊状石莲花
Echeveria gibbiflora 'Caruncutata'

为淡云一个较为特别的栽培品种，其貌不怎样，但奇特，还是为栽培者所爱好。

A special cultivar of Echeveria gibbflora (Hybrid). Their peculiar appearance are faovred by plant enthusiasts.

阔叶石莲花
Echeveria eurychlamys (Diels) A. Berger

为一种较为常见盆栽石莲花品种，亦可露地栽种，但在珠三角也能露地栽种就不知道，原产美洲。

A common potted ornamental from genus Echeveria; could be cultivated outdoor. However, it is not sure whether they could be cultivated outdoor in the Pearl River Delta region; native to America.

吉毕紫石莲
Echeveria gibbiflora DC.

植株多年生草本，多为单生，不易分枝，叶淡紫色，往叶心渐变为紫红色，植株呈莲座生长。

Perennial herbs; mostly solitary, hard to branch; leaves pale purple and grow in a rosette.

铁石莲花
Echeveria gibbiflora DC. (Echeveria metallica Lem.)

叶色似氧化铁棕红色，故此得名。

They are named by its synonym—Echeveria metallica Lem. due to their brownish red leaves, which is the color of iron oxide.

淡云
Echeveria gibbiflora DC. (Hybrid)

原产中美洲，为一种广泛被引种栽培石莲花种类。

A widely introduced cultivar from genus Echeveria; native to Central America.

短柄石莲
Echeveria gracilis (Hybrid)

尖红毛石莲
Echeveria gracilis Rose ex E. Walther

植株翠绿色，叶尖有红点，呈长匙状莲座生长，叶上有短白色绒毛。
Whole plants emerald; leaves long spoon shaped, covered with short white tomentum, grow in a rosette, leaf apex red.

红石莲
Echeveria laui 'Red'(Hybrid)

为一栽培种。
A cultivar of Echeberia.

扇贝石莲花
Echeveria pulvinata Rose

波浪形叶缘似扇贝壳，其叶色灰绿也十分协调，是一种较珍贵的种类。
Echeveria pulvinata Rose are relatively rare species. Their leaves are greyish green and the wavelike leaf margin resembles fan shells.

绒毛石莲
Echeveria pulvinata (Hybrid)

为一栽培种。
A cultivar of Echeberia pulvinata.

特玉莲
Echeveria runyonii 'Topsy-Lurvy'

为劳氏石莲花一个栽培品种，叶形的确较为特别。
A cultivar of Echeveria runyonii with a peculiar leaf shape.

劳氏石莲花
Echeveria runyonii Rose

为一丛生石莲花品种，较为常见小型盆栽观叶花卉，原产中美洲。
Common small potted ornamental from genus Echeveria; usually clustered; native to Central America.

尖红石莲花
Echeveria sp.

为石莲花一个栽培品种，原产美洲南部。
A cultivar of Echeveria; native to southern America.

长序石莲花
Echeveria sp.

本种未鉴定，故不知其正确学名，根据其花序很长很直，取其中文名。

It is an undetermined species. The Chinese name is given here according to its.

匙叶石莲
Echeveria sp.

植株为多年生草本，呈莲座状生长，叶背淡黄绿色，叶面为蓝绿色，匙状肉质叶，栽培较容易，为石莲属一个栽培变种。

A cultivar from genus Echeveria; perennial herbs; grow in a rosette; leaves succulent and spoon shaped, surface blue-green, back pale yellowish green; easy to cultivate.

红花石莲
Echeveria sp.

未找到准确命名，只能看到它的花红色暂给它命名。

A variant from genus Echeveria with red flower. The precise name of this variant has not been found.

灰白石莲花
Echeveria sp.

为石莲花一个栽培品种，适合盆栽。

A intergeneric hybrid cultivar between the genera Echeveria and Crassula, suitable for potting.

红边匙叶石莲
Echeveria sp. (Hybrid)

为一栽培种。

A cultivar of Echeveria.

红边翡翠石莲花
Echeveria sp. (Hybrid)

为一栽培种。
A cultivar of Echeveria.

尖红绒毛石莲
Echeveria sp. (Hybrid)

为一栽培种。
A cultivar of Echeveria.

紫红大石莲

Echeveria sp. (Hybrid)

为一栽培种。

A cultivar of Echeveria.

石莲花和青锁龙杂交种

Echeveria sp. × Crassula sp.

大叶紫石莲
Echevria sp.

未找到正确名。
The precise name of this variant has not been found.

醉美人
Graptopetalum amethystinum (Rose) E. Walther (Hybrid)

肥厚翠绿色的肉质叶，衬托鲜红的叶缘的会联想起饮醉了酒的美人。
The thick and emerald succulent leaves and bright-red leaf margin of Graptopetalum amethystinum easily remind us of a slightly tipsy beauty.

黄玉莲
Graptopetalum sp.

为风车草与石莲花杂交种，叶色半透明和肉质肥厚叶片形态十分可爱，似黄玉。

An intergeneric hybrid between Graptopetalum and Echeveria; leaves fleshy, succulent and translucent.

红边棒槌玉莲
Graptopetalum amethystinum 'Red'

为风车草属棒槌玉莲一个栽培品种。
A cultivar of Graptopetalum amethystinum.

风车草和石莲花杂交种
Graptopetalum sp. × Echeveria sp.

艳美人
Graptopetalum sp.

植株为多年生草本，多分枝，但盆栽环境则不易分枝，肉质叶，肥厚，叶缘有红边。呈莲座生长，形态美观为栽培者喜爱。

Perennial herbs; multi-branched, but not easy to branch in potting condition; leaves succulent and fleshy, leaf margin red, grow in a rosette; particularly beautiful and deeply favored by plants enthusiasts.

金叶仙女之舞
Kalanchoe beharensis Drake (cultivar)

亦称金叶哈伽蓝菜，为一栽培品种，形态确实有点仙女起舞姿态，原产于马达加斯加。

A cultivar; native to Madagascar; look like a dancing fairy.

大叶落地生根
Kalanchoe delagoensis Eckl. & Zeyh. (Kalanchoe tubiflora Raym.-Hamet)

较为常见的一种的伽蓝菜属种类，较耐旱耐湿，植株可成小乔木，其名为落地生根不甚妥，因为在叶缘生出小苗未落地已经生根了。

A common species from the genus Kalanchoe; tolerant to drought and wet; could grow into small arbor.

硃莲
Kalanchoe longiflora V. Coccinec

叶肉质，向上生长，在强光下呈深红色，非常美丽，深受栽培者喜爱。

Leaves succulent, grow upward; display a beautiful dark red color under sunlight; deeply favored by plant enthusiasts.

彩唐印
Kalanchoe tetraphylla H. Perrier (Kalanchoe thyrsiflora Harv.) (Hybrid)

为一栽培种。

A cultivar of Kalanchoe.

长叶彩唐印

Kalanchoe tetraphylla H. Perrier (Kalanchoe thyrsiflora Harv.) (Hybrid)

为一栽培种。

A cultivar of Kalanchoe.

铲叶唐印

Kalanchoe tetraphylla H. Perrier (Kalanchoe thyrsiflora Harv.) (Hybrid)

为一栽培种。

A cultivar of Kalanchoe.

家种彩唐印

Kalanchoe tetraphylla H. Perrier (Kalanchoe thyrsiflora Harv.) (Hybrid)

为一栽培种。

A cultivar of Kalanchoe.

扭叶彩唐印

Kalanchoe tetraphylla H. Perrier (Kalanchoe thyrsiflora Harv.) (Hybrid)

为伽蓝菜属唐印一个栽培变种，翠绿阔匙形肉质叶略带红边，颇吸引人。

A cultivar of Kalanchoe thyrsifoliathyrsiflora from genus Kalanchoe; leaves broad, spoon shapedbroad spatulate, andfleshy, emerald with slightly red leaf margin.

红边耳叶唐印

Kalanchoe tetraphylla H. Perrier (Kalanchoe thyrsiflora Harv.) (Hybrid)

为一栽培种。
A cultivar of Kalanchoe.

洋吊钟

Kalanchoe waldheimii Raym-Hanet & H. Perrier

多见为陆地成片栽种，呈群生。
Usually clustered and cultivated outdoor.

桃美人
Pachyphytum aviferum (Hybrid)

为多叶草属一个栽培变种，叶小球形桃红色，十分别致，一种理想小型盆栽种类。

A cultivar of Pachyphylum genus, leaves small, spherical and peach red; suitable for cultivation.

多叶厚叶草
Pachyphytum compactum Rose

肥厚肉质叶，灰绿色呈紧密莲座生长，很有观赏价值，为适合家庭小型盆栽的种类。
Leaves fleshy, succulent, greyish green and tightly arrange in a rosette; high ornamental value; suitable for home potting.

红艳星美人

Pachyphytum oviferum Purpus

为厚叶草属一个栽培品种，名符其实，每片叶确实够肥厚。

A cultivar from genus Pachyphytum; leaves fleshy.

塔形厚叶莲

Pachyphytum sp. (Hybrid)

为一栽培种。

A cultivar of Pachyphytum.

景天美人
Pachyphytum var. (Hybrid)

多年生肉质草本，为多叶草属栽培变种，叶小球形桃红色叶根偏蓝，为颇受欢迎的观叶花卉。

Perennial succulent herb, cultivar of Pachyphylum genus; leaves small, spherical and peach red, leaf root blue; deeply favored by plant enthusiasts.

黄丽
Sedum adolphii Raym.-Hamet

植株黄绿色，较肥厚肉质叶，莲座生长。

Whole plants yellowish green; leaves succulent and fleshy, grow in a rosette.

莲座光亮景天
Sedum lucidum 'Rosulata'

植株多年生草本，叶呈莲座生长，叶黄翠绿色，叶缘有红边，叶面有光泽。
Perennial herbs; leaves yellowish green, grow in a rosette, leaf margin red, surface sheeny.

高加索景天
Sedum lucidum R. T. Clause (Hybrid)

植株群生，叶深绿色，可能是光亮景天的一个杂交品种。
A hybrid of Sedum lueiduin R. T. clause; clustered; leaves dark green.

光亮景天

Sedum lucidum R. T. Clausen

其肉质叶面较光亮，可能得此名，原产美洲干旱地区，现已被大量引种。

Native to arid areas of America and widely introduced all around the world; leaves succulent and sheeny.

小妞妞

Sedum sp.

植株浅蓝绿色，形似倒泪珠状，故亦称天使的眼泪，是一种适合小型盆栽的观叶花卉。

Whole plants pale blue-green; inverted teardrop shaped, also called the Angel's Tears; a suitable small potted ornamental.

垂叶观音莲

Sempervivum tectorum (Hybrid)

为观音莲栽培变种，叶肉质翠绿，较长，垂下生长，叶尖枣红或紫红色十分优雅。

A cultivar of Se. tectorum; leaves succulent and emerald, long and grow downwards, leaf opex claret or purplish red.

长叶观音莲

Sempervivum tectorum (Hybrid)

为观音莲栽培变种，叶肉质翠绿色，叶尖紫黑色，呈放射莲座生长，形态扁平。

A cultivar of Se. tectorum; leaves succulent and emerald, leaf opex purplish black, grow in a rosette.

观音莲
Sempervivum tectorum (Hybrid)

多年生草本，叶翠绿色，叶尖枣红或紫红色肉质，呈莲座生长。观音莲有多个栽培变种，形态颇受栽种者喜爱

Perennial herb, leaves succulent and emerald, leaf opex claret or purplish red; grow in a rosette; deeply favored by plant enthusiasts.

球状观音莲（菠萝球观音莲）
Sempervivum tectorum L.

植株形态特别像菠萝形状，翠绿色，较稀有的栽培品种。

A rare cultivar; whole plant pineapple shaped and emerald.

番杏科 (Aizoaceae)

番杏科植物虽然在我国无自然分布，但其奇特的形态十分具有观赏价值，已经大量为我国植物公园和爱好者引种、栽种。番杏科原产于南非和纳米比亚，具有独特生态环境，干旱、昼夜温差大、半沙漠和岩石裂缝中。全科共有100多属，已经鉴定的有2000多个种。该科为多年生肉质草本或矮小灌木，花单生雏菊形。在我国不宜露地栽种，只能在具有温度调节的温室内栽种。

Although the plants of family Aizoaceae are not naturally distributed in China, they are widely introduced by plants enthusiasts and parks due to their particular morphologies. They are native to South Africa and Namibia and grow in unique environment, such as rock cracks and semi-desert. This family have about 100 genera and 2000 species, most of them are perennial succulent herbs or small shrubs with solitary and daisy like flowers. They are unsuitable for outdoor cultivation in China and could only be cultivated in greenhouse.

一、菱鲛属（Aloinopsis）

该属共有 10~15 种，均为植株矮小、丛生的肉质草木，肥大块根，肉质茎、叶分不清，花雏菊形，秋末午后或傍晚开花，不耐寒，需在 10 ℃ 以上生长，适宜阳光充足温暖环境下栽种，但不能温度高，冬季要保土壤干燥，早春可扦插或播种繁殖。

The genus Aloinopsis includes about 10 to 15 species, and all of them are small, clustered, succulent herbs with fleshy tuberous roots and stem. Flowers daisy like and usually bloom in the afternoon or evening of late autumn; intolerant to cold weather, suitable for cultivation in sunny and warm condition; keep the soil dry in winter; propagate by cuttage or seeding in the early spring.

粒状菱鲛

Aloinopsis luckhoffii (L. Bolus) L. Bolus

马哈比菱鲛 - 长生草

Aloinopsis malherbei-aizoaceae

莲座菱鲛

Aloinopsis rosulata (Kensit) Schwantes

莲座菱鲛

Aloinopsis rosulata 'Hybrid'

莲座菱鲛

Aloinopsis rosulata var. secdrevolver. Comg

威勒菱鲛

Aloinopsis rosuluta 'Hybrid'

贝毛菱鲛

Aloinopsis sotifera

红菱鲛

Aloinopsis sp.

马哈比菱鲛

Aloinopsis sp.

二、肉堆花属 （Conophytum）

该属有290余种，植株矮小，呈球形或倒圆锥形，生长较缓慢，为多年生肉质草本，花从植株顶部裂缝长出，小雏菊形，花谢后老植株逐渐萎缩成叶鞘，夏末再从叶鞘中长出新植株。不耐寒，应在10℃以上生长，喜低湿温暖环境，夏季环境不能温度太高和多湿，冬季可进行分株或播种繁殖。

The genus Conophytum includes more than 290 species, most of them are slow growing, small, short, spherical or inverted conical perennial succulent herbs. Flowers daisy like and bloom out from the cracks on top of the plant; shrink into sheath after blossom fall, and regrow in the late summer; intolerant to cold weather, suitable for cultivation in warm and low humidity condition; propagate by suckering or seeding in winter.

烛台肉堆花
Cohophytum obcordellu 'Hybrid'

少将
Conophytum bilobum

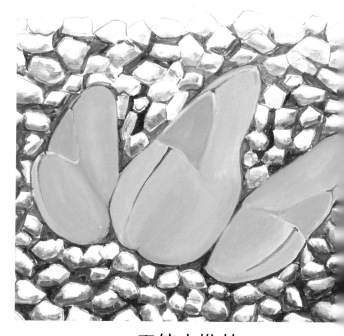

天使肉堆花
Conophytum bilobum (Marloth) N. E. Br.

碧玉肉堆花
Conophytum ectypum N. E. Br.

小球肉堆花
Conophytum minutum (Haw.) N. E. Br.

翠光玉
Conophytum obcordellum (Haw.) N. E. Br.

紫花肉堆花
Conophytum subfenestratum Schwantes

红花肉堆花
Conophytum sp.

三、棒叶花属 （Fenestraria）

该属只有 1~2 种，但花色多样，植株非常矮小，为密集成群生长的多年生肉质草本，分布于纳米比亚半沙漠地区。叶小棒状直立光滑，花雏菊形，白色、淡黄色或淡橙色，夏末至秋末开花，相对较能耐低温，但也不能低于 7℃ 生长。生长期可适度浇水，冬季休眠期应保持干燥，秋季和春季可播种繁殖，春季和夏季可分株繁殖。

The genus Fenestraria includes only 1 to 2 species. All of them are small, short, and compactly clustered perennial succulent herbs; distributed in the semi-desert areas of Namibia; leaves smooth, erect and clavate; flowers white, light yellow or light orange, daisy like; bloom from late summer to late autumn; relatively tolerant to cold weather, but cannot grow below 7 ℃; water moderately during growing period, and keep dry in winter; propagate by seeding in spring or autumn and by suckering in spring or summer.

橙黄花棒叶花
Fenestraria aurantiaca

橙黄花棒叶花
Fenestraria aurantiaca N. E. Br.

白花棒叶花
Fenestraria aurantiaca N. E. Br.

白花棒叶花

Fenestraria aurantiaca N. E. Br.

百花光亮棒叶花

Fenestraria phopalophlla

浅黄花光亮棒叶花

Fenestraria aurantiaca N. E. Br.

棒叶花（橙红花）

Fenestraria sp.

紫花花棒叶花

Fenestraria sp.

四、日中花属 (Lampranthus)

多年生草本，多分枝，呈匍匐状，种植多年会木质化，叶在干燥强光照射下会变红，喜阳光，要求土壤排水良好，春季或秋季可以播种，茎扦插繁殖，原产南非。

Perennial herbs, multi-branched and procumbent; become Lignified after years of cultivation; leaves red under arid and sufficient sunlight condition; need abundant sunlight and good drainage of soil to grow healthily; propagation by stem cutting in spring or autumn; native to South Africa.

红花日中花
Lampranthus aurantiacus Schwantes

弯刀日中花
Lampranthus falciformis (Haw.) N. E. Br.

日中花

Lampranthus sp.

五、生石花属（Lithops）

该属约 40 余种，为无茎多年生肉质草本，分布于纳米比亚和南非的岩缝中和半沙漠地区。植株肥厚，柔软的根状茎，有一堆球果状肉质叶，中央有裂缝，花从裂缝开出，雏菊形，盛夏到中秋开花，不耐寒，生长要控制在 12℃ 以上，喜温暖和阳光充足，初夏至秋末可充分浇水，春季或初夏可播种繁殖，分株繁殖可在初夏进行。该属是引种较多的属，俗称"石头"。

The genus Lithops includes about 40 species, most of them are acaulescent perennial succulent herbs; grow in the rock cracks or semi-desert areas of Namibia and South Africa; whole plants fleshy with soft rhizome; leaves succulent, strobiliform, and crack in the middle; from mid-summer to mid-autumn, daisy like flowers bloom out from the cracks; intolerant to cold weather, suitable for cultivation in warm and sunny condition; water sufficiently from early summer to late autumn; propagate by seeding in spring and by seeding or suckering in early summer.

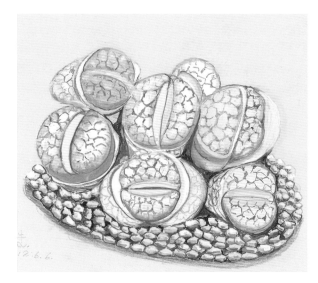

日轮玉
Lithops aucampiae L. Bolus

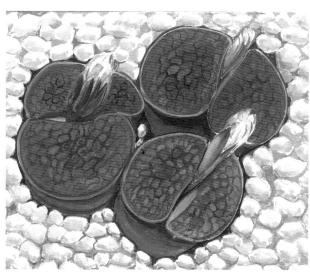

日轮玉
Lithops aucampiae L. Bolus

橙黄花石生花
Lithops francisci N. E. Br.

黄花石头
Lithops gracilidelineata Dinter

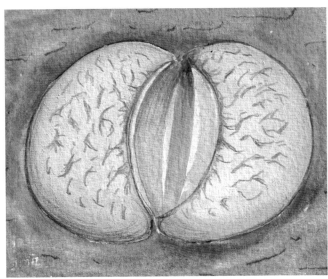

斑绿石生花
Lithops herrei L. Bolus

青玉
Lithops karasmontana N. E. Br.

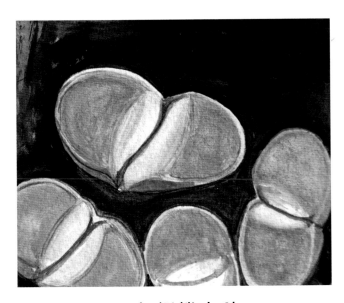

红橄榄变种
Lithops olivacea L. Bolus

红橄榄
Lithops olivacea 'Red Olive'

红花石头
Lithops olivacea 'Rose of Texas'

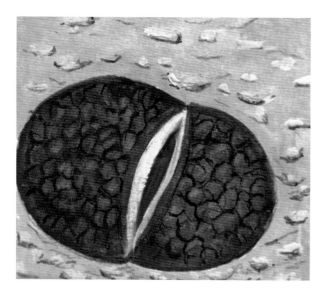

红橄榄变种

Lithops olivacea var. nebrownii D. T. Cole

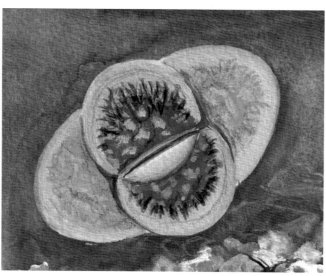

莫高妮石生花

Lithops optica 'Maculate form'

红大内玉

Lithops optica 'Rubra'

紫勋（红大内玉）

Lithops optica 'Rubra'

假截形石生花

Lithops pseudotruncatella 'alpina' Cole

富贵玉

Lithops pseudotruncatella var. dendritica de Boer & Boom

黄花石生花

Lithops ruschiorum N. E. Br.

留蝶玉

Lithops ruschiorum N. E. Br.

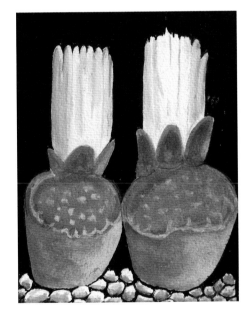

白花石生花

Lithops salicola L. Bolus

福来玉

Lithops sp.

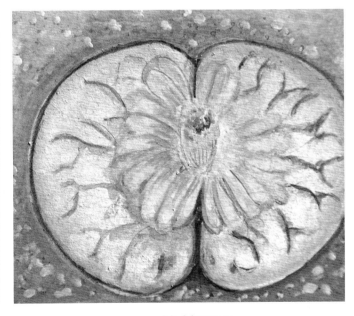

黄花石头

Lithops sp.

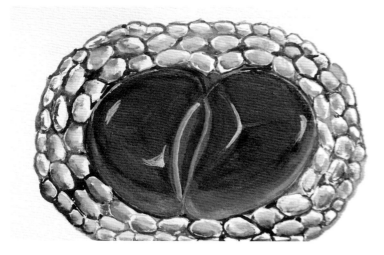

蜡质石生花

Lithops sp.

黄花微纹玉

Lithops sp.

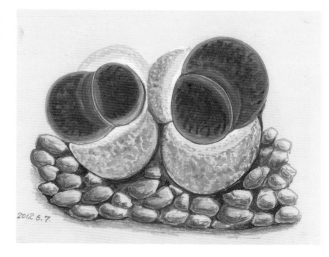

巧克力石头

Lithops sp.

微纹玉

Lithops sp.

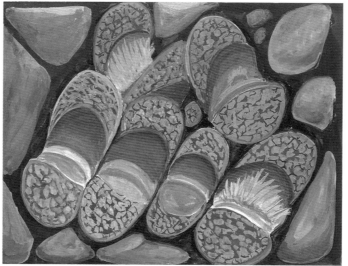

紫翠玉

Lithops sp.

红花石头变种

Lithops verruculosa Nel

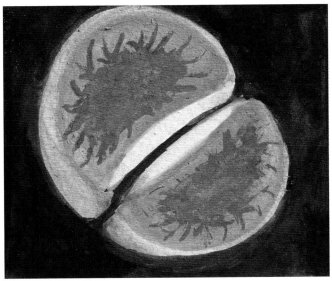

绿玉
Lithops verruculosa Nel

疣突石生花变种
Lithops verruculosa var. glabra de Boer

金王花石生花
Lithops werneri Schwantes & H. Jacobsen

六、对叶生花属（Pleiospilos）

该属有 35 种，植株单生或群生，多年生肉质草本，分布于南非干旱地区。植株呈肥厚肉质的元宝状，叶端三角形或卵形，具有不同颜色的小圆点，花雏菊形，黄色或橙色，夏末至秋初开花，不耐寒，冬季需 12℃以上生长，喜温暖阳光充足环境，初夏至秋初可适度浇水，其余时期应保持干燥，春末至夏初可播种或分株繁殖。

The genus Pleiospilos include about 35 species, most of them are solitary or clustered perennial succulent herbs; distribute in the arid areas of South Africa; whole plants are fleshy and look like shoe-shaped gold ingots; leaf apex triangular or ovoid, with colorful little spots; flowers yellow or orange, daisy like; intolerant to cold weather, suitable for cultivation in warm and sunny condition; water moderately from early summer to early autumn, keep dry in other seasons; propagate by suckering or seeding in later spring and early summer.

波路氏对叶花
Pleiospilos bolusii (Hook. f.) N. E. Br.

亲鸾
Pleiospilos magnipunctatus Schwantes

绿帝玉
Pleiospilos nelii Schwantes

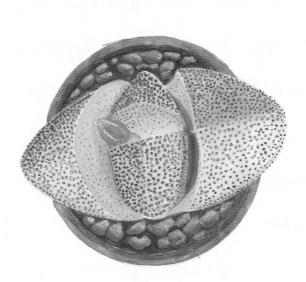

绿帝玉
Pleiospilos nelii Schwantes

绿帝玉变种
Pleiospilos nelii Schwantes

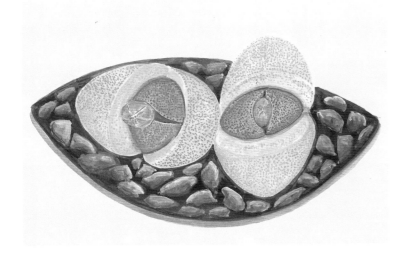

白帝玉
Pleiospilos nelii Schwantes

紫帝玉（变种）
Pleiospilos nelii Schwantes

灰帝玉
Pleiospilos nelii Schwantes

奈尔对叶花
Pleiospilos nelii Schwantes

青鸾
Pleiospilos simulans N. E. Br.

对叶肉堆花
Pleiospilos sp.

红花对叶花
Pleiospilos sp.

七、快刀乱麻属（Rhombophyllum）

该属有 3 种，但花色各异。原分布于南非丘陵边缘地区，叶对生，有 1~2 个短肉质齿，植株为肉质、线状、半圆柱状或镰刀状，花雏菊形，花色浅黄至金黄，夏天白日开花，较耐寒，但也不能低于 7 ℃。喜温暖低湿和阳光充足环境生长，夏天可适度浇水，春季可播种、扦插、分株繁殖。

The genus Rhombophyllum includes 3 species with different flower colors; native to the hilly areas of South Africa; leaves opposite; whole plants succulent, linear, semi-cylindrical or falcate; flowers daisy like, from light yellow to golden; bloom in the summer daytime; relatively tolerant to cold weather, but not lower than 7 ℃; suitable for cultivation in sunny and low humidity condition; water moderately in summer; propagate by seeding, cuttage or suckering in spring.

快刀乱麻
Rhombophyllum dolabriforme Schwantes

尼尔快刀乱麻
Rhombophyllum nelii Schwantes

尼尔快刀乱麻
Rhombophyllum nelii Schwantes

斧状快刀乱麻
Rhombophyllum sp.

斧状叶快刀乱麻

Rhombophyllum sp.

快刀乱麻

Rhombophyllum sp.

八、天女属（Titanopsis）

该属约 5 种，植株具矮短茎，为多年生肉质草本，分布于纳米比亚和南非的半沙漠地区，叶匙状至三角形，肥厚肉质，表面多有疣突呈莲座，花单生雏菊形，黄色至橙色，夏末至初秋开花，不耐寒，栽种时不能低于 10℃。喜温暖低湿和阳光充足的环境，怕高温多湿，春季至夏末可适度浇水，适合在碱性土壤栽种，春季或夏初适合播种繁殖。

The genus Titanopsis includes about 5 species, all of them are perennial succulent herbs with short stems; distribute in the semi-desert areas of Namibia and South Africa; leaves succulent and fleshy, spatulate or triangular, arrange in rosette, leaf surface verrucose; flowers solitary, daisy like, from yellow to orange; bloom from late summer to early autumn; intolerant to cold weather, suitable for cultivation in sunny and low humidity condition; water moderately from spring to late summer, suitable for cultivation in alkaline soils; propagate by seeding in spring or early summer.

长距天女变种

Titamopsis calcarea (Marloth) Schwantes

长距天女变种

Titamopsis calcarea (Marloth) Schwantes

长距天女（红色）

Titamopsis calcarea (Marloth) Schwantes (Red)

长距天女变种
Titamopsis calcarea (Marloth) Schwantes

长距天女
Titamopsis calcarea 'Yellow'

白点天女
Titamopsis primosii L. Bolus

红花天女
Titamopsis primosii L. Bolus

后语 / AFTERWORD

本图谱能够与读者在猴年上半年见面和鉴赏，有赖于各方面的鼎力支持和赞助，特别是天赐三和公司徐金富董事长的全力赞助和支持，使出版关键问题得到完满解决，在此表示十分感谢。另外高旭华高总、徐丹小姐徐总也在各方面给予了大力支持和赞助，在此一并表示感谢。出版社王斌先生以及参与本图谱出版的有关人员，还有在描临过程中给予指点和帮助的关照根先生也表示谢意。总之，对参与本图谱出版工作的有关人士再次表示诚挚的谢意！

朱亮锋
2016.2

Finally, with the great help and sponsorships from everyone around me, these illustrations could be presented to you readers in the first half of the Year of Monkey. Herein, I would like to offer my special thanks to Mr. Xu Jinfu, president of Tian Ci San He corporation, who has supported and sponsored me greatly, and helped me with the publication issues. I would also like to thank Mr. Gao Xuhua and Miss Xu Dan for their great supports and sponsorships. Additionally, I am truly grateful for the assistances from Mr. Wang Bin and others who have assisted me with the publication procedure. And I have been deeply appreciated that Mr. Guan Zhaogen have offered me advises and assistances during my hand-painting process. All in all, I offer my deepest and most sincere thankfulness to everyone who have participated in the creation and publication of these illustrations!

Zhu Liangfeng
Feb. 2016

主要参考书

1. 艾里希·葛茨·格茨，格哈德·格律纳，威利·库尔曼. 2007. 仙人掌大全. 辽宁科学技术出版社
2. 朱亮锋. 2011. 自然珍藏图鉴丛书多肉植物. 南方日报出版社
3. Edger and Brian Lamb. 1969. Rocket Encyclopedia of CACTI. Macmillan Publishing CO. N.Y.
4. Rod 8 Ken Preson-Mafham. CACTI. The Illusteated Dictionary.
5. 朱亮锋. 1983. 肉质植物. 直接传媒（国际）
6. 朱亮锋. 2009. 仙人掌. 大公报出版有限公司

拉丁名索引
Index to Scientific Names

A

Acanthocalycium spiniflorum	3
Adromischus cristatus	221
Adromischus sp.	221
Acanthocalycium spiniflorum	3
Aeonium	233
Aeonium	233
Aeonium (Hybrid)	222
Aeonium 'Greenovia diplocgcla'	231
Aeonium arbereum (Hybrid)	223
Aeonium arboreum (Hybrid)	224
Aeonium arboreum 'Atroprupureum' (Hybrid)	229
Aeonium arboreum 'Atropurpureum' (Hybrid)	224
Aeonium arboreum 'Atropurpureum' (Hybrid)	225
Aeonium arboreum 'Atropurpureum' (Hybrid)	226
Aeonium arboreum 'Atropurpureum' (Hybrid)	227
Aeonium arboreum 'Hybrid'	228
Aeonium 'Greenovia diplocgcla'	230
Aeonium diplocyclum	231
Aeonium haworthii	232
Aeonium tabulitorme	232
Aloe 'Hybrid'	207
Aloe aculeata 'Red'	161
Aloe aculeata 'Variegata'	161
Aloe aculeata 'Variegata' (Red)	162
Aloe aculeate 'Variegata'	162
Aloe africana	164
Aloe africana × A. ferox	163
Aloe africana × marlothii (Hybrid)	163
Aloe alooides	165
Aloe barberiae	165
Aloe barteri	166
Aloe berhana	166
Aloe boylei	167
Aloe brevifolia	167
Aloe brownii	168
Aloe candelabrum	168
Aloe castanea	169
Aloe ciliaris	169
Aloe comosa	170
Aloe comptonii	170
Aloe conifera	171
Aloe crassicaulis	171
Aloe cryptopoda	172
Aloe dabenorisana	172
Aloe dewetii	173
Aloe distans	173
Aloe divaricata	174
Aloe dyeri	174
Aloe echeveria	175
Aloe excelsa	175
Aloe ferox	177
Aloe ferox × marlothii (Hybrid)	176
Aloe glauca	178
Aloe globuligemma	178
Aloe hardyi	179
Aloe hexapetala	179
Aloe 'Hybrid'	207
Aloe ibitiensis	180
Aloe khamiesensis	180
Aloe khamiesensis	181
Aloe lineata	181
Aloe littoralis	182
Aloe longibracteata	182
Aloe lutescens	183
Aloe macroclada	183
Aloe maculata	184
Aloe marlothii	186
Aloe marlothii	187
Aloe marlothii × ferox (Hybrid)	185
Aloe melanacatha	187
Aloe mutabilis	188
Aloe peglerae	189
Aloe percrassa	190
Aloe pillansii	191
Aloe plicatilis	192
Aloe pluridens	193
Aloe pluridens × lineata (Hybrid)	192
Aloe polyphlla	193
Aloe pretoriensis	194
Aloe reitzii	195
Aloe rupestris	196
Aloe saponaria	184
Aloe somaliensis	197
Aloe soutpansbergensis	197
Aloe sp.	208
Aloe sp.	209
Aloe speciosa	179
Aloe speciosa × A ferox	198
Aloe spicata	198
Aloe striata Haw.	199
Aloe succotrina	199
Aloe suprafoliata	200
Aloe thraskii	201
Aloe unfoloziensis	202
Aloe vaombe	202
Aloe variegata	203
Aloe volkensii × ferox (Hybrid)	204
Aloe volkensii × ferox (Hybrid)	204
Aloe volkensii	205
Aloe vryheidensis	206
Aloe wickensii	207
Aloinopsis luckhoffii	277
Aloinopsis malherbei-aizoaceae	277
Aloinopsis rosulata	278
Aloinopsis rosulata 'Hybrid'	278
Aloinopsis rosulata var. secdrevolver	278
Aloinopsis rosuluta 'Hybrid'	278
Aloinopsis sotifera	279
Aloinopsis sp.	279
Aloinopsis sp.	281
Ariocarpus fissuratus	4
Ariocarpus furfuraceus	4
Ariocarpus kotschoubeyanus	5
Ariocarpus retusus	6
Ariocarpus retusus subsp. trigonus	5
Ariocarpus scaphirostris	7
Arrojadoa rhodantha	7
Arrojadoa rhodantha	8
Astrophytum asterias	9
Astrophytum asterias var. nudum	8
Astrophytum capricorne	10
Astrophytum myriostigma	11
Astrophytum myriostigma 'Variegata'	11
Astrophytum myriostigma 'Variegata'	10
Astrophytum myriostigma var. columnarie	12
Astrophytum myriostigma var. nudum	13
Astrophytum ornatum	13
Austrocactus patagonicus	14

B

Blossfeldia liliputana	14
Buiningia brevicylindrica	15

C

Carnegiea gigantea	16
Cephalocereus senilis	17
Cleistocactus jujuyensis	17
Cleistocactus straussii	18
Cohophytum obcordellu 'Hybrid'	280

301

Conophytum bilobum 280	Echereria laui 246	Echinocereus sp. 34	Epiphyllum 'Red Bird' (Hybrid) 133
Conophytum ectypum 281	Echeveria (Hybrid) 246	Echinocereus subinermis 34	Epiphyllum 'Red' (Hybrid) 134
Conophytum minutum 281	Echeveria 'Yallum' (Hybrid) 247	Echinocereus subinermis var. ochoterenae 35	Epiphyllum crenatum 47
Conophytum obcordellum 281	Echeveria agavoides (Hybrid) 247		Epiphyllum floribundum 47
Conophytum sp. 281	Echeveria chihuahuaensis 248	Echinocereus triglochidiatus 35	Epiphyllum oxypetalum 48
Conophytum subfenestratum 281	Echeveria elegans 248	Echinocereus triglochidiatus 36	Epithelantha micromeris 49
Corryocactus squarrosus 18	Echeveria eurychlamys 249	Echinocereus triglochidiatus var. melanaeanthus 36	Escobaria minima 49
Coryledon macrantha 'Red Variegata' 234	Echeveria gibbiflora 250		Escobaria strobiliformis 50
	Echeveria gibbiflora (Hybrid) 251	Echinocereus triglochidiatus var. paucispinus 37	Espostoa huanucoensis 50
Coryphantha calipensis 19	Echeveria gibbiflora 'Caruncutata' ... 249		Espostoa lanata 51
Coryphantha elephantidens 19	Echeveria gracilis 252	Echinocereus triglochiidiatus var. Melanacanthus 37	Espostoa lanata 52
Coryphantha pallida 20	Echeveria gracilis (Hybrid) 251		Eulychnia breviflora 53
Coryphantha poselgeriana 20	Echeveria laui 'Red'(Hybrid) 252	Echinocereus viridiflorus 38	Eulychnia ritteri 53
Coryphantha robustispina 21	Echeveria pulvinata (Hybrid) 253	Echinocereus websterianus 38	**F**
Coryphantha werdermannii 22	Echeveria pulvinata 253	Echinofossulocactus kellerianus 40	Fenestraria aurantiaca 282
Cotyledon macrantha 234	Echeveria runyonii 'Topsy-Lurvy' 254	Echinofossulocactus multicostatus 40	Fenestraria aurantiaca 283
Cotyledon undulata 235	Echeveria runyonii 254	Echinofossulocactus multicostatus 'Variegata' 41	Fenestraria phopalophlla 283
Cotyledon undulata 236	Echeveria sp. 255		Fenestraria sp. 283
Cotyledon undulata var. haematophyllas .. 237	Echeveria sp. 256	Echinofossulocactus pentacanthus 'Variegata' 41	Ferocactus coloratus 54
	Echeveria sp. 257		Ferocactus cylindraceus 54
Cotyledon undulata 'Hybrid' 235	Echeveria sp. (Hybrid) 258	Echinofossulocactus phyllacanthus ... 42	Ferocactus emoryi 55
Crassula 'Gollum' 237	Echeveria sp. (Hybrid) 259	Echinofossulocactus vaupelianus 42	Ferocactus gracilis 55
Crassula 'Morgan's Beauty' 238	Echeveria sp. (Hybrid) 257	Echinofossulocactus zacatecasensis 'Variegata' 43	Ferocactus hamatacanthus 56
Crassula arborescens 239	Echeveria sp. (Hybrid) 258		Ferocactus herrerae 57
Crassula arborescens 'Haematophyllas' 238	Echeveria sp. × Crassula sp. 259	Echinofossulocactus zacatecasensis 'Variegata' 43	Ferocactus histrix 57
	Echevria sp. 260		Ferocactus histrix 58
Crassula barklyi 239	Echinocactus grusonii 24	Echinofossulocacus albatus 39	Ferocactus latispinus 58
Crassula capitella 241	Echinocactus johnsonii 25	Echinopsis (Hybrid) 44	Ferocactus lindsayi 59
Crassula capitella 'Flammeus' 240	Echinocactus platyacanthus 25	Echinopsis candicans 44	Ferocactus pilosus 59
Crassula capitella 'Red' 240	Echinocactus texensis 26	Echinopsis deserticola 44	Ferocactus pottsii 60
Crassula deceptor 241	Echinocereus engelmannii 26	Echinopsis formosa 45	Ferocactus rectispinus 61
Crassula falcata 242	Echinocereus fendleri 27	Echinopsis obrepanda 45	Ferocactus reppenhagii 61
Crassula falcata 242	Echinocereus ferreianus 27	Echinopsis oxygona 46	Ferocactus robustus 62
Crassula perfoliata var. falcata 242	Echinocereus ferreirianus var. lindsayi .. 28	Echinopsis smrziana 46	Ferocactus schwarzii 62
Crassula perfoliata var. falcata 243	Echinocereus ledingii 28	Epiphyllum (Hybrid) 130	Ferocactus sp. 63
Crassula sp. 243	Echinocereus longisetus 29	Epiphyllum (Hybrid) 131	Ferocactus townsendianus 64
Crassula sp. 245	Echinocereus palmeri 30	Epiphyllum (Hybrid) 132	Ferocactus townsendianus 63
Crassula sp. 244	Echinocereus pectinatus 31	Epiphyllum 'America Sweetheart' (Hybrid) 134	Ferocactus wislizeni 65
D	Echinocereus pulchellus 32		Ferocactus wislizeni 64
Discocactus horstii 22	Echinocereus rigidissimus 32	Epiphyllum 'America Sweetheart' (Hybrid) 135	Frailea castanea 66
Discocactus horstii 23	Echinocereus rigidissimus subsp. rubispinus 33		**G**
Disocactus eichlamii 24		Epiphyllum 'Frau H. Wegener' (Hybrid) 132	Gasteria carinata 210
Dudleya pulverulenta 245	Echinocereus scheeri 33		Gasteria carinata 'Variegata' 211
E	Echinocereus schmollii 39	Epiphyllum 'Mae Marsh' (Hybrid) 133	Gasteria disticha 211

Gasteria disticha 212	**H**	Lithops optica 'Maculate form' 288	Melocactus bellavistensis 103
Gasteria disticha 'Vartegata' 212	Haageocereus multangularis 82	Lithops optica 'Rubra' 288	Melocactus bellavistensis 104
Gasteria excavate 210	Hatiora (Hybrid) 83	Lithops pseudotruncatella 'alpina' ... 288	Melocactus bellavistensis subsp. onyc-hacanthus
Gasteria excavate 'Hybrid' 213	Hatiora 'Grande' 84	Lithops pseudotruncatella var. dendritica....	 103
Gasteria sp. 213	Hatiora epiphylloides 84	 289	Melocactus caesius 105
Gasteria sp. 214	Haworthia attenuata 214	Lithops ruschiorum 289	Melocactus concinus 104
Glanodulicactus uncinatus 66	Haworthia attenuata var. radula 215	Lithops salicola 289	Melocactus ernestii 105
Graptopetalum amethystinum 260	Haworthia fasciata 'Elegans' 215	Lithops sp. 289	Melocactus matanzanus 106
Graptopetalum amethystinum 'Red' ... 261	Haworthia herbacea 'Variegata' 216	Lithops sp. 290	Melocactus oreas 106
Graptopetalum sp. 261	Haworthia limifolia 216	Lithops verruculosa Nel 290	Melocactus peruvianus 107
Graptopetalum sp. 262	Haworthia marginata 217	Lithops verruculosa 291	Melocactus salvadorensis 107
Graptopetalum sp. × Echeveria sp. 262	Haworthia radula 215	Lithops verruculosa var. glabra 291	Monvillea spegazzinii 108
Crassula falcata J. C. Wendl. 243	Haworthia reinwardtii 217	Lithops werneri 291	**N**
Greenovia diplocycla Webb ex Bolle 231	Haworthia sp. 218	Lobivia aurea 87	Neobuxbaumia euphorbioides 108
Gymocalycium dnudatum 70	Haworthia viscosa 'Variegata' 219	Lobivia aurea var. fallax 88	Neolloydia conoidea 109
Gymnocactus subterraneus 67	Heliocereus speciosus 85	Lobivia bruchii 88	Neoporteria napina 109
Gymnocalycium anisitsii 67	Hildewintera × Echinopsis 86	Lobivia caineana 89	Neoporteria sp. 110
Gymnocalycium baldianum 68	Hylocereus undatus 86	Lobivia densispina 89	Notocactus magnificus 110
Gymnocalycium bruchii 68	**K**	Lobivia einsteinii 90	Notocactus magnificus 111
Gymnocalycium buenekeri 69	Kalanchoe beharensis 263	Lobivia famatimensis 91	Notocactus muricatus 111
Gymnocalycium cardenasianum 69	Kalanchoe delagoensis 263	Lobivia hertrichiana 91	Notocactus rauschii 112
Gymnocalycium mibanovichii var. friedrichii 71	Kalanchoe longiflora 264	Lobivia jajoiana 92	Notocactus uebelmannianus 112
Gymnocalycium mihanovichii 71	Kalanchoe tetraphylla 264	Lobivia maximiliana subsp. caespitosa	Notocactus vanvlietii 113
Gymnocalycium mihanovichii 76	Kalanchoe tetraphylla 265	 93	Notocatus Hybrid 113
Gymnocalycium mihanovichii var. friedrichii 72	Kalanchoe tetraphylla 267	Lobivia pentlandii 93	Notocatus Hybrid 114
Gymnocalycium mihanovichii var. friedrichii 73	Kalanchoe tetraphylla 266	Lobivia schieliana var. leptacantha 94	Notocatus sp. 114
Gymnocalycium mihanovichii var. friedrichii 74	Kalanchoe tubiflora Raym.-Hamet 263	Lobivia silvestrii 94	Notocatus sp. 115
Gymnocalycium mihanovichii var. stenogonum 75	Kalanchoe tetraphylla 266	Lobivia silvestrii 95	**O**
Gymnocalycium mihanovichii var. stenosonun var. friedrichii 75	Kalanchoe tetraphylla 265	Lobivia winterina 95	Obregonia denegrii 116
Gymnocalycium nigriareolatum 76	Kalanchoe waldheimii 267	Lobivia wrightiana 96	Opuntia bergeriana 117
Gymnocalycium quehlianum 77	**L**	**M**	Opuntia bigelovii 117
Gymnocalycium saglionis 78	Lampranthus aurantiacus 284	Mammillaria backebergiana 96	Opuntia elatior 116
Gymnocalycium saglionis 79	Lampranthus falciformis 284	Mammillaria casoi 97	Opuntia gosseliniana 118
Gymnocalycium saglionis 80	Lampranthus sp. 285	Mammillaria elegans 97	Opuntia longiareolata 118
Gymnocalycium stellatum 80	Leuchtenbergia principis 87	Mammillaria hubertller 98	Opuntia phaeacantha 119
Gymnocalycium tillianum 81	Lithops aucampiae 286	Mammillaria longimamma 98	Opuntia robusta 119
Gymnocalycium uruguayense 81	Lithops francisci 286	Mammillaria parkinsonii 99	Opuntia rufida 120
Gymnocalycium vatheri 82	Lithops gracilidelineata 287	Mammillaria senilis 99	Opuntia salmiana 121
	Lithops herrei 287	Mammillaria theresae 100	Opuntia sp. (Andes, Peru) 157
	Lithops karasmontana 287	Matucana aurantiasa 100	Opuntia sp. (Andes, Peru) 158
	Lithops olivacea 287	Matucana pallarensis 101	Oreocereus doelzianus 122
	Lithops olivacea 'Red Olive' 287	Melocactus albicephalus 101	Oreocereus hendriksenianus 122
	Lithops olivacea 'Rose of Texas' 287	Melocactus azureus 102	Oreocereus neocelsianus 123
	Lithops olivacea var. nebrownii 288	Melocactus bahiensis 102	**P**

Pachycereus schottii 123	Pleiospilos sp. 294	Selenicereus grandiflorus 142	schwarzii 152
Pachyphytum aviferum 268	Pterocactus fischeri 136	Selenicereus pteranthus 143	Turbinicarpus subterraneus 153
Pachyphytum compactum 268	Pterocactus reticulatus 137	Sempervivum tectorum 275	**W**
Pachyphytum oviferum 269	Pterocactus tuberosus 137	Sempervivum tectorum (Hybrid) 274	Weberocereus biolleyi 154
Pachyphytum sp. (Hybrid) 269	Puya raimondii 159	Sempervivum tectorum (Hybrid) 273	Weingartia kargliana 154
Pachyphytum var. (Hybrid) .. 270	**R**	Sempervivum tectorum (Hybrid) 275	Weingartia lanata 155
Parodia buiningii 124	Rebutia aureiflora 138	Setiechinopsis mirabilis 143	Weingartia westii 155
Parodia crassigibba 125	Rebutia torquata 138	Srombocactus disciformis 144	Wilcoxia schmollii 37
Parodia echinopsoides 125	Rebutia violaciflora 139	Stenocereus hystrix 144	Wilcoxia schmollii 39
Parodia horstii 126	Rhombophyllum dolabriforme 295	Stetsonia coryne 145	Wilcoxia viperina 156
Parodia leninghausii 126	Rhombophyllum nelii 295	Sulcorebutia steinbachii 145	**Z**
Parodia maassii 127	Rhombophyllum nelii 296	**T**	Zygocactus truncatus 156
Parodia maassii var. subterranea 127	Rhombophyllum sp. 296	Tacinaga inamoeua 146	
Parodia neoarechavaletae 128	Rhombophyllum sp. 297	Thelocactus bicolor 147	
Parodia schumanniana 128	**S**	Thelocactus heterochromus ... 148	
Parodia sp. 124	Schlumbergera 'Delicatum' (Hybrid) 140	Thelocactus hexaedrophorus . 148	
Peniocereus viperinus 129	Schlumbergera 'Goldcharm' (Hybrid) 140	Thelocactus tulensis 149	
Peniocereus viperinus var. ... 129	Schlumbergera 'Le Vesuv' (Hybrid) 141	Titamopsis calcarea 298	
Pereskia bahiensis 130	Schlumbergera Hybrid 141	Titamopsis calcarea 299	
Phyllocactus 'Verana' 131	Schlumbergera truncata 142	Titamopsis calcarea (Red) 317	
Pilosocereus leucocephalus .. 135	Sedum adolphii 270	Titamopsis calcarea 'Yellow' . 299	
Pilosocereus pachycladus 136	Sedum lucidum 272	Titamopsis primosii 299	
Pleiospilos bolusii 292	Sedum lucidum 271	Trichocereus 'Variegata' (Hybrid) 149	
Pleiospilos magnipunctatus .. 292	Sedum lucidum 'Rosulata' 271	Turbinicarpus laui 150	
Pleiospilos nelii 293	Sedum sp. 272	Turbinicarpus pseudopectinatus 150	
Pleiospilos nelii 294		Turbinicarpus pseudopectinatus 151	
Pleiospilos simulans 294		Turbinicarpus schmiedickeanus subsp.	